TAMING CANCER

Taming Cancer

21st Century Biology and

the Future of Cancer Medicine

Drew N. Kelner, Ph.D.

BioCentury Press

All Rights Reserved
No part of this book may be used or reproduced in any manner whatsoever without written permission except for brief quotations embodied in critical articles and reviews.

Copyright©2024 Drew Nathaniel Kelner, Ph.D.

Cover designed by Ethan S. Kelner

Figures 1, 2, 14, 15, and the cover photograph of dividing cancer cells were licensed from dreamstime.com. Except for Figure 13, which was licensed from Cell Press, all other figures were licensed from shutterstock.com.

ISBN: 979-8-9886081-0-3

BioCentury Press

tamingcancerbook@gmail.com

A Subsidiary of Shenandoah Biotechnology Consulting, LLC

Charlottesville, VA

Printed in the United States of America

To

Deb, the love of my life, 33 years and counting,
and to our beloved sons, Noah and Ethan, our greatest achievements.

Table of Contents

List of Illustrations ix

List of Acronyms xi

Author's Note and Acknowledgements xiii

Introduction xv

1. In Search of the Magic Bullet 1

2. The Plastic, Fantastic Cell 25

3. The Caretaker and the Gatekeeper 37

4. Deciphering the Source Code 65

5. Living in an RNA World 91

6. Genes Gone Wild 103

7. Chaos by Design 129

8. The Ultimate Sugar High 141

9. The Wound That Does Not Heal 159

10. The Seed and the Soil..................................171

11. Hard to Kill 185

12. Targeting Tumors the 21st Century Way 199

13. Living Drugs 223

14. The 21st Century Magic Bullets 241

15. The Future of Cancer Medicine 259

Notes 281

Index 319

About the Author..335

List of Illustrations

Figure 1. Activation of B cells by antigens 11

Figure 2. The structure of human IgG 14

Figure 3. Schematic diagram of hematopoiesis 33

Figure 4. Formation of the peptide bond 42

Figure 5. The superoxide radical 48

Figure 6. Phases of the cell cycle 53

Figure 7. The structure of DNA in the human cell 66

Figure 8. The Central Dogma of Molecular Biology 70

Figure 9. The structure of the ribosome 74

Figure 10. The structure of tRNA 75

Figure 11. The Philadelphia Chromosome 121

Figure 12. The structure of ATP 142

Figure 13. The Hallmarks of Cancer 181

Figure 14. Co-stimulatory interactions of T Cells 234

Figure 15. An antibody and a CAR-T construct 235

List of Acronyms

ACT	adoptive cell transfer
ADC	antibody-drug conjugate
ALL	acute lymphoblastic leukemia
APC	antigen-presenting cell
ATP	adenosine triphosphate
CAR-T	chimeric antigen receptor T (cell)
CLL	chronic lymphocytic leukemia
CML	chronic myelogenous leukemia
CR	complete regression (complete response)
CRS	cytokine release syndrome
CSC	cancer stem cell
ctDNA	circulating tumor DNA
CTLA-4	cytotoxic T-lymphocyte-associated protein 4
DDR	DNA damage response
DNA	deoxyribonucleic acid
DTC	disseminated tumor cell
EBV	Epstein-Barr virus
ECM	extracellular matrix
EGFR	epidermal growth factor receptor
EMT	epithelial-to-mesenchymal transition
ENCODE	Encyclopedia of DNA Elements
ESC	embryonic stem cell
FMT	fetal microbiota transplant
HGP	Human Genome Project
HIF	hypoxia-inducible factor
HLA	histocompatibility locus antigen
HPV	human papilloma virus

HSC	hematopoietic stem cell
MDSC	myeloid-derived stromal cell
MET	mesenchymal-to-epithelial transition
MHC	Major Histocompatibility Complex
MMP	matrix metalloproteinase
mRNA	messenger RNA
mtDNA	mitochondrial DNA
NCI	National Cancer Institute
NET	neutrophil extracellular trap
NIH	National Institutes of Health
NK	natural killer (cell)
ORR	objective response rate
OS	overall survival
PCR	polymerase chain reaction
PD-1	programmed death-1
RISC	RNA Inducible Silencing Complex
RNA	ribonucleic acid
rRNA	ribosomal RNA
ROS	reactive oxygen species
RSV	Rous sarcoma virus
scFv	single-chain variable fragment
siRNA	small interfering RNA
SNP	single nucleotide polymorphism
TCA	trichloroacetic acid (cycle)
TCR	T cell receptor
TIL	tumor-infiltrating lymphocyte
TME	tumor microenvironment
tRNA	transfer RNA
TVEC	Talimogene laherparepvec

Author's Note and Acknowledgements

When I set out on my long journey of writing *Taming Cancer*, I hoped the book would speak to a general audience by explaining scientific concepts in everyday language while also serving as a resource for those who want to delve deeper into the science. In this way, the book could be of value to anyone interested in understanding cancer at a conceptual level while offering a path for further exploration of the most complex and fascinating human ailment by physicians, scientists, nurses, students, and other interested parties. To serve the latter objective, the Notes contain references to scientific papers with additional information on key scientific concepts. My goal was to impart to the reader a semblance of the enthusiasm and wonder I have experienced throughout my life-long love affair with biology and medicine.

There are many people I need to thank for their help and support. First, to the most important people: the true love of my life, my brilliant and talented wife, Deb, and our incredible young men, Ethan and Noah. I could not have done this without your love, patience, and encouragement. I am indebted to Noah for the stimulating discussions about evolution and human biology, and to Ethan for his creative talents in the design of the book cover and valuable marketing advice. I am thankful beyond measure to Dr. Shawn Eisenberg, the Vice President of Technical Operations at Acelyrin, for his critical review of the manuscript. Shawn's eagle-eyed attention to detail, insightful questions, and remarkable ability to discover hidden errors (and even a few outright mistakes) made this a much better book. I am also grateful to the group of friends who willingly served as beta-readers as I endeavored to finalize the contents. Obviously, any remaining errors are on me.

Finally, I must thank all the wonderful colleagues and friends I've had the pleasure of working with over the past 35 years. This book is a tribute to you and all the other dedicated professionals in the biopharmaceutical industry who design, develop, and manufacture innovative medicines that help alleviate the human suffering caused by our most devastating diseases.

Drew N. Kelner, Ph.D.
Charlottesville, VA
April 20, 2024

Introduction

My interest in cancer goes back at least half a century.

As a teenager in the seventies, I recall looking at the stunning graphics in *Scientific American* that showed how cancer cells invade neighboring tissues, leaving a path of cellular destruction that portends an excruciating dance with death. I remember thinking that this terrifying aberration of our biology is part of who we are. I often wondered how our cells turn against us and why this devastating illness exists. I also remember thinking that I wanted to write about cancer one day.

Here we are, more than 50 years later. During a sizable chunk of that time, I had the privilege of participating in the development of medicines for the treatment of cancer, along with drugs for other grievous diseases, including hemophilia, anemia, osteoporosis, and rheumatoid arthritis. These medicines, called biopharmaceuticals (or biological drugs), are different from familiar drugs like Nexium and Lipitor, which are synthesized by combining chemicals in large vats to create pills for oral use by patients.

Biopharmaceuticals are manufactured in cell culture systems using equipment similar to what you might see at your local brew pub. Think animal cells instead of yeast (they are not all that different), and, instead of beer, a turbid solution, generally pale red in color, that contains, amongst the many living cells, precious biopharmaceutical proteins manufactured by those cells.

The names of some of these drugs might be familiar to some readers. Enbrel® and Humira®, which are competing medicines for rheumatoid arthritis and other autoimmune diseases, are used by millions of people around the world. These therapeutics cannot be administered orally because they are delicate protein molecules that would degrade in the highly acidic environment of the digestive tract. Rather, biopharmaceuticals are injected into patients, either by the intravenous route (into the vein), in the muscle (intra-muscular route), or under the skin (subcutaneously).

Currently, cancer therapeutics comprise the lion's share of the biopharmaceutical drugs in development as we continue our lengthy battle with this most challenging and perplexing of human diseases, the "Emperor of All Maladies," as Dr. Siddhartha Mukherjee christened it in his scintillating 2010 biography of cancer. Decades of research have demonstrated that cancer is, from a biological standpoint, extraordinarily complex at the cellular and molecular levels. Moreover, every cancer is a "one-of-a-kind" affair, developed under a unique set of circumstances such that each tumor has an inimitable molecular fingerprint. The distinctive genetic, biochemical, and biological nature of cancer is also why achieving long-term clinical success is so difficult.

Now, more than ever, there is reason for hope in our battle with this terrifying disease. We live in a time when powerful new approaches show significant potential in the fight against cancer. Over the past few decades, scientists have assembled a comprehensive (but incomplete) scientific understanding of the molecular machinations of the human cell. As a result, we can now map, in exquisite detail, the aberrant molecular circuitry that drives the destructive growth of cancer cells.

From such a molecular understanding, the underlying biochemical defects permissive to tumor growth are being elucidated. The acquisition of this knowledge raises the possibility that the molecular switches that allow this life-threatening disease to spread throughout the body in the deadly process of metastasis can be turned off or, at the very least, controlled sufficiently to improve both the duration and the quality of life of cancer patients.

As a result of these insights into the nature of cancer, a dramatic shift is underway from the cancer therapeutic triad of surgery, radiation, and chemotherapy— "cut, burn, and poison"—to exciting new molecular approaches that harness the power of biotechnology to exploit the weaknesses of cancer cells. These developments include immunotherapeutics, new medicines that can stimulate the human immune system to seek out and destroy tumor cells that have escaped the continuous process of immune surveillance that guards us against disease.

Introduction

By confronting cancer with biomolecules that can curtail its growth, it is now possible to realistically imagine a world in which a diagnosis of metastatic cancer is no longer, by default, the existential threat it represents today. Rather, the experience will be akin to that of patients with medically manageable chronic conditions such as diabetes and rheumatoid arthritis. While this malicious malady inherent to our biology cannot be eradicated, perhaps, at last, it can be tamed.

Over the past decade, clinical evidence has emerged that the new medical tools described in this book are capable, in small subsets of fortunate patients with metastatic cancer, of achieving long-term remissions and even, on occasion, eradicating detectable cancer cells. Not long ago, the ability to develop and introduce therapeutic agents into clinical use that specifically and effectively target human cancer would have been found only in science fiction. In this century, the likelihood of breakthrough treatments for the most feared affliction of our time has never been more promising.

This is the age of molecular medicine—some have called it the "Bio-Century"—and its wonders await.

Chapter 1

In Search of the Magic Bullet

The idea that we carry an innate ability to resist disease dates to antiquity. The ancient Greeks observed that even in the face of a plague (likely smallpox or typhus) that decimated Athens in the fifth century B.C.E., some of the afflicted recovered and remained protected from the fatal effects of the deadly disease for years. Thucydides, the fifth century B.C.E. author of the definitive text on the Peloponnesian War, noted that these lucky individuals were protected from suffering the full measure of the disease: "The bodies of dying men lay one upon the other... [But] those who had recovered from the disease ... had now no fear for themselves; for the same man was never attacked twice—never at least fatally."[1]

The basis of microbial infection remained unknown until the middle of the nineteenth century, a product of the landmark achievements of two European scientists, German physician Robert Koch and French microbiologist Louis Pasteur. These towering figures in the history of biology showed that the great infectious diseases of the age were caused by specific types of microorganisms that live in the vast sea of invisible life surrounding us.

Koch's work on tuberculosis, anthrax, and cholera laid the foundation for our understanding of infectious diseases. According to his findings, encapsulated in what came to be known as *Koch's Postulates*, proof of infection with a specific microbial agent can be demonstrated if an organism isolated from an infected individual can be grown in the laboratory and subsequently shown to cause the same disease when introduced into an uninfected recipient.[2] If these conditions are met, a relationship between the infectious agent and the disease it causes is unequivocally established.

Pasteur's breakthrough vaccinations for rabies, diphtheria, and anthrax in the late nineteenth century demonstrated the power of vaccination as a preventative agent against diseases caused by microorganisms. These achievements were based on the pioneering discoveries of British physician Edward Jenner nearly a century before.

Jenner had heard for decades that milkmaids bearing cowpox lesions on their hands and legs rarely contracted smallpox. Since cowpox caused a disease that is highly similar—but far milder in its effects—when compared to the more deadly smallpox, and some people exposed to smallpox remained free from its ravages, there had to be a natural capability to prevent the disease from taking hold in certain individuals. In a remarkable (and risky) human experiment in 1796, Jenner took some scrapings from cowpox lesions he found on the hands and legs of a milk maiden named Sarah Nelms. Next, he placed the cowpox scrapings under the skin of an 8-year-old child named John Phipps, his gardener's son.[3]

This phase of the experiment mimicked a procedure called *variolation*, in which scrapings from smallpox lesions were placed under the recipients' skin to prevent the disease. Turkish traders introduced variolation into Europe in the early 1700s and had practiced it in Asia for centuries. While the practice reduced the incidence of smallpox in the population, variolation infected about 2-3% of its recipients with smallpox, sometimes fatally.[4]

Jenner's key idea was that it might be possible to protect against smallpox infection using material from the related disease, cowpox, without the risk of transmitting deadly smallpox to the recipients. Two months after young Master Phipps was inoculated with the cowpox-lesion-derived material, Jenner proceeded with the riskiest part of his experiment; he challenged the child by inoculating him with material from a fresh smallpox lesion. Amazingly, there were no ill effects at all. John Phipps did not even suffer the usual fever and malaise that routinely followed variolation.

This astonishing discovery happened at the end of the eighteenth century, more than a century-and-a-half before the discovery of antibodies, the powerful molecules of immunity that help protect us from disease. From this astounding result, Jenner confirmed his hypothesis: a small amount of diseased material can stimulate a protective response. With this enormous leap forward in preventing one of the deadliest infectious agents in the history of humankind, the science of vaccination was born.

Early in the twentieth century, medical science had advanced sufficiently to explore the biological and chemical bases of host immunity to infectious diseases. Many questions remained: How could we have immunity to a limitless set of substances (called *antigens*) in our environment that can elicit an immune response? How could we generate protective responses for such a large array of potential irritants?

These questions were the focus of the work of German physician Paul Ehrlich. Ehrlich was born in 1854 in Upper Silesia, Germany, in the southwest corner of modern-day Poland. Educated as a medical doctor, he recognized that the identification of the cell as the unit of biological life by German scientists **Matthias Schleiden and Theodor Schwann** in the middle of the nineteenth century had moved biology's central axis from the level of the whole organism in the nineteenth century to the level of the cell in the twentieth.

Ehrlich realized that to understand biology, we needed a way to look inside the cell to discover the secrets of the biological molecules responsible for cellular functions. He believed that the microscope, which allowed biologists to view cellular structures, but not the molecules that comprise them, had taken biology about as far as it could go in the quest to understand the living chemistry at the heart of cellular functions.

The German physician called the various biological processes in the cell the "*partial cell functions.*" He noted that "for a further penetration into the important, all-governing problem of cell life even the most highly refined optical aids will be of no use to us."[5] Thus, he issued a call for more sophisticated analytical instrumentation that would not come to fruition until after the Second World War, almost half a century later.

Recognizing that a true understanding of biological processes required that investigations go beyond micro-anatomical descriptions to the underlying chemical mechanisms at play, Ehrlich noted, "Since what happens in the cell is chiefly of a chemical nature and since the configuration of chemical structures lies beyond the limits of the eye's perception we shall have to find other methods of investigation for this."[6]

These scientific insights were remarkably predictive of the future of biological and medical science. Ehrlich demonstrated a penchant for prescience when he proclaimed, "This approach is not only of significant importance for a real understanding of the life processes, but also the basis for a truly rational use of medicinal substances."[7] Herein lay his key insight: truly effective medicines must target specific biological processes rather than merely provide relief from symptoms. By understanding how medicines work—by investigating what scientists now call the *mechanism of action*—the drug development process can be guided by biological knowledge rather than by trial and error. This approach, Ehrlich realized, would require a detailed understanding of the biochemistry of the cell. Succinctly put, he proclaimed, "We have to learn to aim chemically."[8]

Ehrlich's work on the neutralization of diphtheria and botulinum toxins by *anti-toxins* in the blood of infected individuals convinced him that the toxin and the anti-toxin must interact in a highly specific way. This specificity, he proposed, was rendered by precise interactions between the toxin and the anti-toxin mediated by what Ehrlich called *"side chains."*

He envisioned these side chains as chemical structures with individualized shapes. When the side chains of a toxin are complementary to those on an anti-toxin—that is, the side chains of one fit together in three-dimensional space with the side chains of the other—the toxin and the anti-toxin will latch onto each other in a firm chemical embrace.

We can think of analogies: a lock and a key or a pair of tessellating tiles that fit perfectly into each other. Ehrlich envisioned that if he could find an anti-toxin that perfectly fits in a specific way with a known toxin, it would be possible to neutralize the toxin.

In this vision, the anti-toxin was envisioned as a "magic bullet"—a specific, precise, and effective means to target a toxic substance in the body, bind to it, and thereby prevent the toxin from causing physiological harm. This was a powerful vision, and it would take decades of research to discover that the anti-toxins—Ehrlich's "magic bullets"—are proteins called *antibodies*. The antibodies, which are made by white blood cells called *B lymphocytes*, comprise only part of the extraordinarily complex system of immunity that protects us from disease.

Ilya Mechnikov was born in 1845 in a small village near Kharkiv, Russia, in modern-day Ukraine. Encouraged to study science by his mother, he was a natural science prodigy who lectured neighborhood children on botany and geology when he was six.[9]

After studying biology at the city's university, Mechnikov collaborated with Russian zoologist Alexander Kovaleskyin—first in Naples, Italy, and then in St. Petersburg, Russia, where the two scientists fled following a cholera outbreak in southern Italy in 1865.[10] Mechnikov completed his doctoral studies in 1867, earning a Ph.D. in embryology.[11]

While pursuing his studies in comparative embryology in 1882, Mechnikov was examining starfish larvae under a microscope. He had chosen the larvae of the genus *Bipinnaria* because they provide an excellent model system for biological study due to a convenient matter of their anatomy. *Bipinnaria* larvae are transparent, making it possible to peer inside them with a microscope and observe the movement of cells.

Mechnikov noticed that cells were moving inside the larvae engulfing particles of food. It occurred to him as he observed the cells engulfing the food particles that these cells might also be involved in protecting the larvae from microbes, microscopic organisms that can cause disease. "These wandering cells in the body of the larva of a starfish, these cells eat food ... but they must eat up microbes too!"[12]

He devised a simple experiment in which he placed tiny thorns inside the larvae to assess whether the wandering cells would react to the presence of foreign substances. As predicted, the cells responded to the foreign bodies in their midst. "He noted that the cells within the larvae were no longer moving around aimlessly, but were instead aggregated around the foreign bodies, as if to drive them out."[13]

Mechnikov called the process in which the wandering cells engulf foreign matter *phagocytosis*, from the Greek words' *phage*, meaning "to eat," and *cyte* (from the Greek *ketos*), meaning "cell." These cells, which he named *phagocytes*, can engulf foreign matter.

A further test of his theory involved placing fungal spores in water fleas of the genus *Daphnia*. Mobile cells in the flea could also engulf the spores. Obviously, these cells played a role in protecting the organism from infection. Further experiments with higher organisms, such as rabbits, convinced him that he had discovered a general mechanism of immunity present in all multi-cellular organisms. Extrapolating to humans, he noted, "Our wandering cells, the white cells of our blood—they must be what protects us from invading germs."[14]

The Russian scientist had found a powerful, innate defense against infection, a means for the body to neutralize potential microbial threats. "Where natural immunity is concerned, and man enjoys this in respect of a large number of diseases, it is a question of the phagocytes being strong enough to absorb and make the infectious microbes harmless."[15]

Shortly after the discovery of phagocytosis, a German scientist named Emil von Behring made another profound discovery. Von Behring had worked directly with Robert Koch, and near Paul Ehrlich, at the Institute for Infectious Diseases in Berlin. He applied that strong scientific foundation to his studies of diphtheria, a bacterial illness that posed a serious and potentially lethal threat to children in the early twentieth century.

Von Behring found that he could remove all the cells from a sample of an infected animal's blood (cell-free blood is called *serum*), infuse the infected animal's serum into the bloodstream of an uninfected animal, and thereby protect the uninfected animal from a challenge with the causative agent of diphtheria, the bacterium *Corynebacterium diphtheria*.[16] This serum transfer experiment demonstrated that a substance in an infected animal's bloodstream could protect against diphtheria infection. Known as an anti-toxin by biologists at the time, the agent, later called an antibody, was (we now know) a protein that can bind to a specific target on a foreign substance in the body—in this case, to a target on the surface of the bacterial cells.

Von Behring received the first Nobel Prize in Physiology or Medicine in 1901 for demonstrating that the immune response was not solely a matter of phagocytic action by Mechnikov's wandering cells. As a result of von Behring's work, a heated debate ensued in the biological community about whether immunity was a matter of cellular activity (phagocytes) or, alternatively, whether anti-toxins (antibodies) in the blood provided protection against microbes.

With Ilya Mechnikov as a major proponent, the former idea was called the cellular basis of immunity. The latter idea, supported by the work of von Behring and Ehrlich, was known as the "humoral" basis of immunity in recognition of the role of blood—one of the humors (bodily fluids) described by Hippocrates—in providing protection against infectious microbial organisms.[17] As it turned out, both sides had equal merit.

In awarding the 1908 Nobel Prize in Physiology or Medicine to Mechnikov and Ehrlich for their groundbreaking work on immunity, the Nobel committee equally recognized the critical importance of both immune mechanisms. This view would be strengthened during the following century of investigation, which clearly showed that the cellular and humoral immunity mechanisms work together in a highly coordinated fashion to regulate the immune response.

The discoveries of Ehrlich, Mechnikov, and von Behring on the nature of immunity launched the science of immunology. Their work revealed the presence of two major subsystems called *innate immunity* and *adaptive immunity*. These two subsystems, the "arms" of the immune system (in common biological vernacular), interact with each other through complex molecular signaling networks to coordinate the overall immune response.

As the name implies, we are born with the elements of innate immunity already in place and on the job. Our skin, the body's largest organ (by surface area), is the primary layer of protection against invasion. Immune cells called *neutrophils* and *macrophages* circulate throughout the body to kill and engulf microbial invaders. Innate immunity is a generalized response triggered by exposure to foreign substances, regardless of their identity or origin. The innate response does not need to develop over time; innate immunity is triggered without requiring previous exposure.

The other arm of immunity, the adaptive immune response, requires (as the name implies) that the system learns over time to distinguish antigens (proteins or chemical substances bound to proteins) originating inside our bodies from those derived from foreign sources. Throughout our lives, the cells of adaptive immunity continually sample the antigens in the body, learning to distinguish foreign antigens from our own and thereby guarding against potential threats that require an immediate response.

In the adaptive arm, the first exposure to a specific antigen, a process that immunologists call *priming*, does not trigger a significant response. Rather, it trains the system to respond to subsequent exposures to the antigen. Once primed, adaptive immunity is ready to respond when re-exposed to the priming antigen. Herein lies the basic principle of vaccination, in which a virus or piece thereof trains the immune system to respond in the event of a future infection by that virus.

Following antigen exposure, the cellular constituents of the adaptive immune system primed by previous antigen exposure go into production mode, generating a humoral (antibody) response to the antigen by antibody-producing white blood cells (B lymphocytes). In addition, an adaptive cellular response is stimulated, characterized by the rapid activation of immune cells called *T cells* (*T lymphocytes*) that are specific for the antigen. These activated T cells can kill foreign cells (for example, bacteria and viruses). Known as the "cellular soldiers" of adaptive immunity, T cells circulate throughout the body following antigen stimulation in search of the foreign antigen that launched them into action.[18]

⸻

Von Behring's experiments conclusively demonstrated that immunity engendered by serum transfer is specific to the organism that caused the disease in the animal from which the serum was taken. Thus, the transfer of serum from an animal infected with diphtheria can protect against a subsequent challenge with the bacteria responsible for diphtheria, but not against the bacteria responsible for botulism (and vice versa).

Ehrlich's side chain theory proposed that this specificity was related to the molecular characteristics of the antigens on the surface of infectious organisms. In turn, the chemical properties of the antigens' side chains provided specific binding sites for the anti-toxins' side chains, which fit snugly in three-dimensional space with specific structural features of the antigens.

Given the observed specificity of the response, and the complexity of this process at the level of molecular structure, how was it possible that the body can recognize and generate a specific response to the millions of antigens present in the environment? Stated in the terms used by modern-day immunologists, what mechanisms are at play in providing the vast repertoire of antibodies that can be elicited by antigen stimulation?

According to Ehrlich's theory, the answer resided in the presence of a limitless array of chemical structures on the surfaces of the anti-toxin-producing cells (later renamed *antibody-producing cells*) in the circulation. Ehrlich reasoned that for these cells to manufacture an anti-toxin for a toxin that is present in the bloodstream, they must have a way of identifying the chemical side chains on circulating toxins. Envisioning a mechanism that might explain how the anti-toxin-producing cell recognizes and responds to the toxin, Ehrlich proposed that the side chains of the molecules on the surfaces of the anti-toxin-producing cells must be the same as those on the anti-toxin produced by that cell.

The rationale for this proposal is as follows: If the side chains on the surface of the cell fit together with those of the toxin, and the anti-toxin made by the cell has the same side chains as those on the toxin-binding structure on the cell surface, then the side chains on the anti-toxin produced by that cell will also fit with those on the toxin. This format provided a ready answer to the question of how the anti-toxin-producing cell creates a molecule with side chains that can bind to the side chains of the toxin amidst the extraordinarily complex biochemical milieu of bodily tissues.

At the time, not only was the biochemical structure of anti-toxins unknown, the identity of antibodies as members of a family of related protein molecules involved in immunity had not yet been established. Ehrlich had no concept of the existence of cell surface proteins on each B lymphocyte that are, in fact, the antibodies produced by that cell. Paul Ehrlich's proposal was, therefore, unadulterated genius.

It is unfathomable to this twenty-first century biochemist how Ehrlich made such a leap beyond what was known at the time. In the complete absence of any data supporting his contention, Ehrlich formulated a prescient hypothesis on the biological basis of toxin/anti-toxin specificity decades before the nature of antigens and antibodies was revealed.

Refinements of Ehrlich's side chain theory did not emerge for half a century. In 1955, British immunologist Niels Jerne proposed that the existence of a vast array of pre-existing antibodies in the serum was responsible for antibody diversity. Once an antibody finds an antigen with which it forms a tight biochemical "fit," Jerne reasoned, the presence of the antibody-antigen complex stimulates the B cell that produced that antibody to divide. He called his idea the *natural selection theory* of antibody production.

The main problem with the natural selection theory as formulated by Jerne was the lack of an explanation for how a B cell can sense when its antibody molecules are bound to antigens in the circulation. One could propose that following antigen binding, a signaling event takes place between the circulating antibodies and the antibody-producing cells. However, there was no evidence for this mechanism, nor was there a conceivable explanation why this might be so.

This mystery was solved shortly thereafter, in 1960, when Australian immunologist Frank Macfarlane Burnet modified Jerne's natural selection theory. Burnet proposed that the antigen-recognizing protein sticking out of the membrane on the surface of the antibody-producing cell *is* the antibody produced by that cell. This idea harkened back to Ehrlich's concept that the anti-toxins with their toxin-specific side chains resided on the cell surface.

Burnet's solution to the question of B cell activation is shown in Figure 1.

B-cell activation

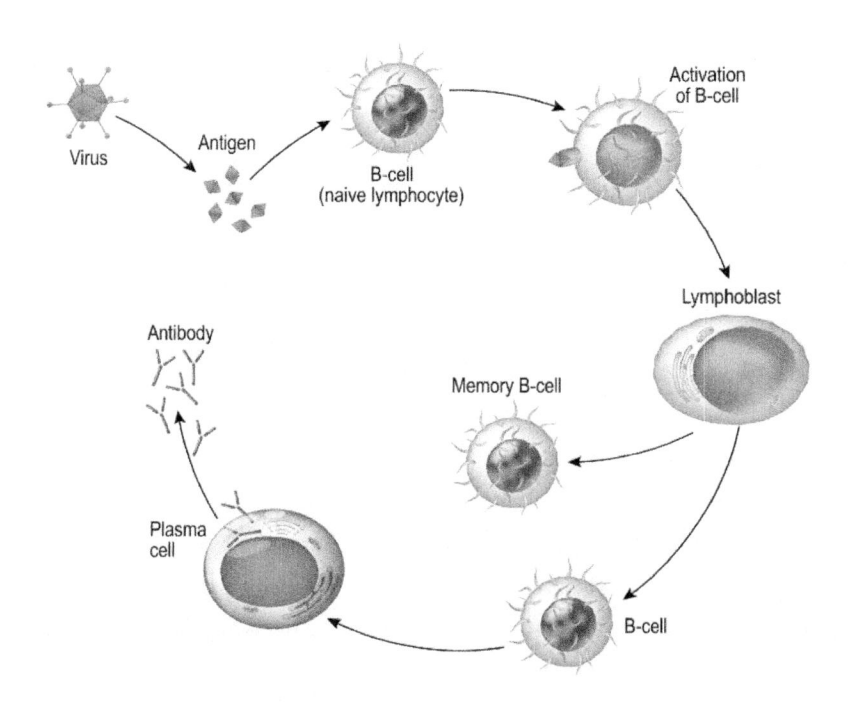

Figure 1. Activation of B cells by antigens

Burnet proposed that once an antigen found an appropriate molecular fit with a surface protein on a B cell (which is the antibody produced by that cell), the cell would be stimulated to divide. This would produce many copies of the B cell that manufacture the same antibody as the originally stimulated B cell. In this way, large numbers of antibody molecules are created and subsequently secreted from the B cells into the circulation. Since this idea was based on the concept that a stimulated B cell can be copied (cloned) to make many copies, Burnet called his proposal the *clonal selection theory* of antibody production.

In addition to antibody-producing B cells (called *plasma cells*), another type of B cell, called a *memory B cell*, is created following antigen exposure. While activated B cell levels decay over time, memory B cells are responsible for long-term immune recognition. These two cell types are derived from the same ancestral activated B cell (called a *lymphoblast*) during B cell activation.

Once formed, memory B cells can be activated upon further antigen exposure, such that they rapidly proliferate and create new antibody-producing plasma cells. Memory B cells are found both in lymph nodes and in the circulation, and they persist long after the cessation of antibody production by plasma cells.[19]

As our knowledge of adaptive immunity accumulated, a mystery remained. How could the genome possibly code for the millions of types of antibodies needed to cover the environment's tremendous diversity of antigens?

The first piece of this puzzle was solved by the mid-1970s with the advent of the *recombinant DNA revolution*. A series of discoveries in molecular biology provided, for the first time in human history, a technique for inserting genes of interest into bacteria, yeast, and mammalian cells. Thus, living cells could be used as factories for manufacturing the proteins encoded by the genes. These new techniques in molecular biology enabled clever investigators to unravel the mechanism of antibody diversity, providing compelling evidence that Burnet's clonal selection theory was correct.

As Burnet predicted, antibody molecules are indeed expressed on the surface of *naïve* (unstimulated) *B cells*, which are then selected for proliferation following binding of an antigen to a surface antibody on the B cell.

As the selected B cell proliferates, the genes responsible for antibody production undergo rearrangements that lead to the refinement and maturation of the antibodies to increase their antigen specificity. In this process, called *affinity maturation*, multiple generations of refined antibody molecules are generated to increase the tendency of the antibody to bind to the antigen that stimulated the response.

With the tools of genetic analysis wrought by recombinant DNA technology—that is, the ability to isolate, amplify (make many copies), and sequence the DNA—the mystery of antibody diversity was unraveled, and the structure of antibodies was elucidated. While some of our proteins are comprised of a single chain of amino acids—a single *polypeptide*, in the language of biochemistry—other proteins are comprised of two or more distinct polypeptide chains. Such is the case for antibodies. Protein analysis work performed in the 1960s showed that *immunoglobulin G (IgG)*, the major antibody type found in human serum, is comprised of four polypeptide chains: two copies each of two different polypeptides called the *heavy chain* and the *light chain*.

IgG (Figure 2) comprises about 80% of the antibody population in human serum. The remaining types (*classes*) of antibodies, called IgM, IgA, IgD, and IgE, each have a distinct architecture, with different numbers and orientations of the polypeptides that form the overall structure. The four polypeptide chains that comprise IgG are bound together in a Y-shaped molecular construct that is connected by bonds (called *disulfide bonds*) between a sulfur atom in the amino acid cysteine on one heavy chain and another sulfur atom in an adjacent cysteine on the other heavy chain (thereby creating a *disulfide bridge*). This region of the antibody, called the *hinge*, provides flexibility around a stable rotational axis. There are four IgG subtypes (*sub-classes*), called IgG1, IgG2, IgG3, and IgG4, each with a characteristic disulfide bond structure.[20]

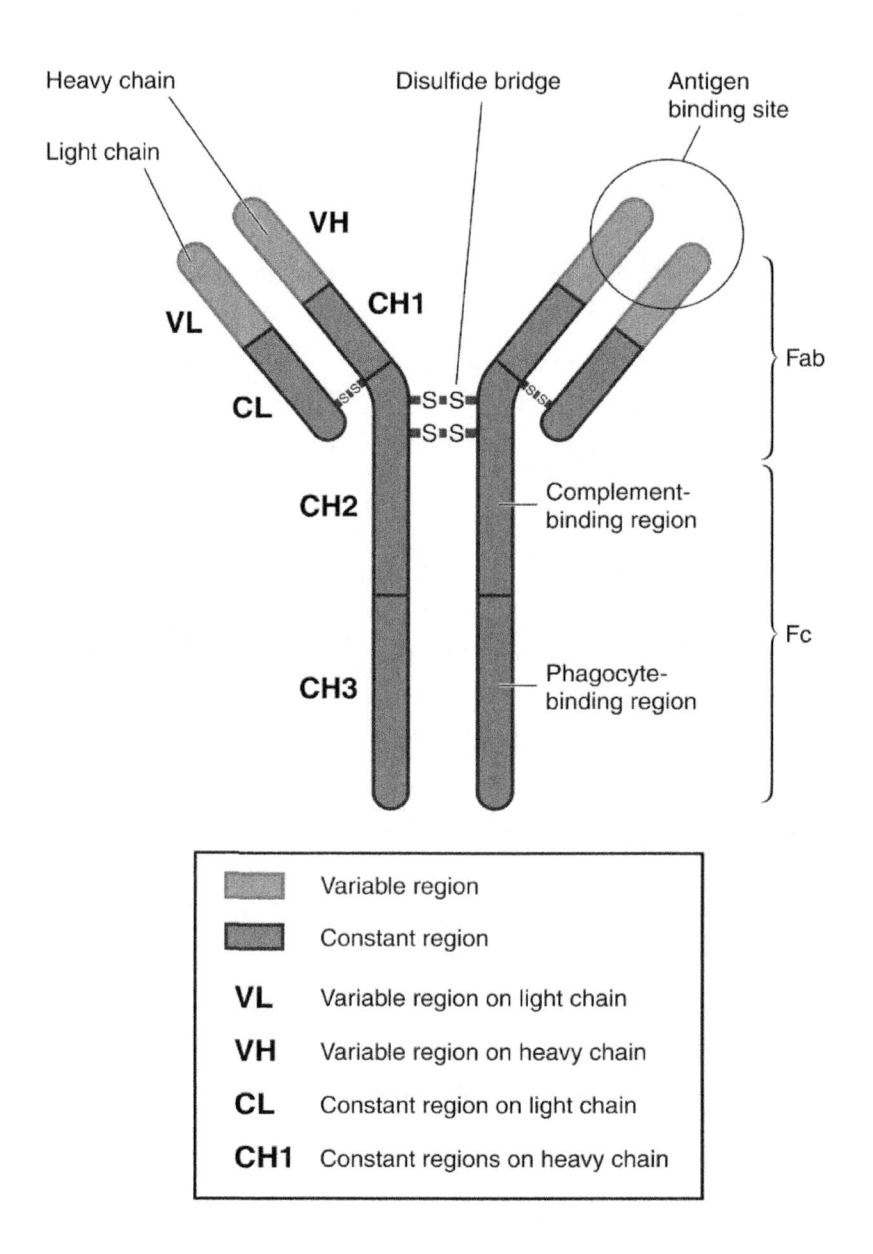

Figure 2. The structure of human IgG

The heavy (H) chain contains both variable (V_H) and constant (C_H) sections (called *domains*); similarly, the light (L) chain has both variable (V_L) and constant (C_L) domains. The Y-shaped IgG structure has two identical antigen binding sites, as shown at the top of Figure 2. These antigen-binding sites, comprised of sequences with significant variability from antibody to antibody, consist of sections of the V_H and V_L domains of the heavy and the light chains, respectively.

If the antibody binds its target antigen with binding sites at the top of the arms of the Y-shaped structure (the sequences above the hinge form the antibody's *Fab region*), what is the function of the sequences below the hinge, known as the *Fc region*?

The Fc region is responsible for stimulating what are known as the *effector functions*. These functions are elicited by the binding of sequences on the Fc region of the antibody to specific proteins on the surfaces of immune cells to stimulate the destruction of a target cell (e.g., a bacterium). As the name implies, the constant regions of the antibody where the effector function binding sites reside (e.g., complement- and phagocyte-binding regions) are similar in sequence from antibody to antibody of the same class (e.g., IgG) and sub-class (i.e., IgG1, IgG2, IgG3, or IgG4). Each IgG sub-class has a distinct effector function profile. In addition to stimulating cell killing by immune cells, an effector function called *FcRn binding* allows IgGs to remain in the circulation for up to about three weeks.[21]

To provide for antibody diversity, the heavy chain and the light chain are assembled from information encoded in multiple immunoglobulin genes. These genes can be "mixed and matched" to create multiple protein sequences from the individual heavy and light chain genes. The biochemical processing of the genetic sequences that code for immunoglobulins is unusually imprecise during the assembly of the various immunoglobulin genes. This provides multiple sequence variants during assembly that increase antibody sequence diversity.

Further antibody sequence diversity is added by the presence, in the antigen-binding sequences of the variable (V_H and V_L) domains, sequences that are prone to rapid mutation (these are called *hypervariable regions*).

The combination of processes described above that contribute to the diversity of antibody sequences provides the capability to confront an unlimited variety of antigens during our lifetimes.

Argentinian biochemist Cesar Milstein was the son of Jewish immigrants from Ukraine who emigrated to South America around the turn of the twentieth century. During his education, Milstein developed an interest in the emerging science of immunology. He was particularly fascinated by the challenge of understanding how such a vast diversity of antibodies is achieved. Years later, he recognized that he had underestimated the complexity of the problem:

> What attracted me to immunology was that the whole thing seemed to revolve around a remarkably simple experiment: take two different antibody molecules and compare their primary sequences. The secret of antibody diversity would emerge from that. Fortunately at the time I was sufficiently ignorant of the subject not to realise how naive I was being.[22]

When he started his work on antibody diversity, Milstein had no idea that beyond the sequences themselves, the biosynthesis of antibodies by the cellular machinery involves diversity-promoting mechanisms that had not even entered the realm of the conceivable in the 1970s. This was a time before the discovery of the process of immunoglobulin gene assembly and its significant role in the formation of mature immunoglobulins from the encoded DNA sequences found in the cell's nucleus.

In the 1970s, a decade after receiving a Ph.D. in biochemistry from the University of Cambridge,[23] Milstein was working with German immunologist George Kohler at the Basel Institute of Immunology studying mutations in the antigen-binding region of antibodies. While growing antibody-producing cells in tissue culture, they ran into an insurmountable barrier: the cells in culture were not growing well, such that they were incapable of producing enough of the antibodies for detailed molecular analysis.

Kohler and Milstein had a brainstorm. They thought of a way to create antibody-producing cells that would grow well in cell culture. An emerging cell culture technique called *cell fusion* made it possible to put two different cell types together in culture and subject them to conditions where they can merge into one, resulting in a hybrid cell with the properties of both cell types. If one could select the correct pair of cell types, it might be possible to generate a hybrid cell that grows well in cell culture and produces significant quantities of the desired antibody.

The researchers began by immunizing mice with the antigen of interest. Knowing that the spleen is loaded with B lymphocytes, they made cellular preparations from the spleens of the immunized mice. The cells from the spleen preparation were exposed in culture to a rapidly growing mouse tumor cell line (called a *myeloma*). Like most tumor cell lines, the myeloma cells grow easily and continuously in culture. Unlike normal cells, which have a limited lifespan in cell culture, tumor cell lines are often immortal when grown in the lab, and therefore can be kept alive and replicating for decades. This unlimited potential for survival and cell division is due to multiple dysfunctions in the biochemical mechanisms that regulate the lifespan of normal cells. These controls over the lifespan of cells ensure that damaged, aged cells are replaced by new, healthy cells.

Using biochemical screening methods, Kohler and Milstein were able to identify cells in the culture derived from the fusion of a spleen-derived B cell and a myeloma cell.[24] Such a cell, called a *hybridoma*, is a hybrid cell that can be grown in culture for extended periods of time while secreting a continuous supply of antibodies due to the presence in the hybridoma of immunoglobulin genes derived from the B cell.[25] As Milstein noted in his Nobel Prize acceptance speech in 1984 (an award he shared with Kohler and Niels Jerne), "The resultant hybrid was an immortal cell capable of expressing the antibody activity of the parental antibody-producing cell, the immortality acquired from the myeloma."[26]

The power of Kohler and Milstein's hybridoma technology rests in the ability of the method to create many copies of an antibody, derived from a single type of B cell, called a *monoclonal antibody*. Such an antibody is specific for a single structural feature of the antigen. While Paul Ehrlich would have called the single structural feature on the antigen that binds to an antibody a side chain, in modern immunological parlance, the small piece of the antigen (which is comprised of about five to seven amino acids) bound by the antibody is known as an *epitope*.

Whereas each hybridoma cell produces one kind of antibody, specific for one epitope—that produced by the B cell that fused with the tumor cell to create the hybridoma—conventional immunization, which involves the collection of the serum from the immunized animals, provides a mixture of different antibodies rather than a single monoclonal antibody population. This is because there are multiple epitopes on an antigen (perhaps hundreds) that generate a variety of antibodies that bind different epitopes on the same antigen. Thus, animal immunization results in the production of what is known as *polyclonal antibodies*, a mixture of antibodies derived from different B cells, each antibody type binding to its specific epitope on the antigen.

The invention of hybridoma technology stands as one of the twentieth century's most significant discoveries in the biological sciences. Hybridoma technology provides a long-term source of antibodies specific for a defined epitope on an antigen. As a result, this method for producing monoclonal antibodies has profoundly impacted biological research, with direct applications in the development of new medicines and diagnostic tests for serious human diseases, including cancer.

Kohler and Milstein did not patent their technology, choosing instead to share it with the world. Ironically, this important discovery was a by-product of their search for the secret to antibody diversity. This spectacular development was not the initial goal of their work. It was not the first, and no doubt far from the last, case where a profound scientific invention came about despite the original experimental intent, and not because of it.

Paul Ehrlich believed that anti-toxins had the potential to provide therapeutic agents for human diseases with the ability to act in a highly specific manner. Because of this specificity, such therapeutic agents could, in theory, be highly effective. In addition, he believed that such agents would have minimal side effects and low toxicity due to the reduction in interactions of the agent with structures that are not targeted by the therapy.

Such off-target effects are a major source of the side effects observed with traditional, synthetic drugs, which have lower molecular specificity for their targets than biological drugs such as monoclonal antibodies. Avoidance of off-target effects is amongst the most compelling reasons why biopharmaceutical drugs—those based on biological molecules, rather than those derived by chemical extraction or synthesis—usually have superior safety profiles relative to traditional medications, which lack the precise targeting inherent to biological drugs.

As a result of Kohler and Milstein's discovery, monoclonal antibody therapeutics have been developed for metabolic diseases, autoimmune diseases (such as rheumatoid arthritis, psoriasis, and Crohn's disease), the prevention of organ transplant rejection, osteoporosis, and many types of cancer. The science of antibody engineering, which uses the tools of modern molecular biology to optimize the structure and function of antibody therapeutics, has been advancing rapidly since the 1990s.

Due to their relative safety and high target specificity, monoclonal antibodies are the most abundant form of recombinant protein used for the treatment of human diseases. Hundreds of antibodies are under evaluation in clinical trials around the world. Many of the most widely used biological agents for the treatment of human diseases are based on monoclonal antibody technology, including a drug called Humira.

This monoclonal antibody, which targets an immune stimulatory molecule called *Tumor Necrosis Factor* (*TNF*), is indicated for the treatment of several auto-immune disorders, including ulcerative colitis, psoriasis, and several forms of arthritis. Humira is currently the most prescribed drug in the world, with annual global sales exceeding $20 billion.

The most widely used monoclonal antibodies in the practice of oncology (cancer medicine) include Herceptin®, Avastin®, and Rituxan®. These medicines are prescribed hundreds of thousands of times yearly for patients suffering from many types of cancer—including cancer of the breast, brain, colon, lung, liver, kidney, ovary, and several types of leukemia and lymphoma. By bringing these three antibodies to the routine practice of medicine, from Rituxan's approval for non-Hodgkin's lymphoma in 1997 to Avastin's approval for colon cancer in 2004, Genentech (now owned by Roche) played a leading role in the introduction of antibody therapeutics into routine clinical practice for the treatment of human cancer.

While cancer strikes people of all ages—including, tragically, children—statistics show that "the incidence of cancer increases exponentially with age."[27] The likelihood of a cancer diagnosis at age 30 is about 70 per 100,000 (.07%). By age sixty, the incidence rate increases about 14-fold, to about 1,000 per 100,000 (1%).[28]

There are several reasons for this dramatic increase in cancer incidence with age. Since cancer is a disease of the genome, the accumulation of genetic damage over time is a significant contributor. Like all cellular functions, aging degrades the cellular quality control systems that ensure the integrity of the genetic material. As a result, the cell's ability to detect and repair damaged DNA becomes less efficient as we age.

The degradation of the cellular machinery responsible for the cell's metabolism is also a major contributor to the increase of cancer incidence with age. Cellular metabolism encompasses all the biochemical reactions that extract the chemical energy contained in our food to provide the energy needed by the cell to drive the myriad biochemical reactions that support cellular functions.

When we suffer an injury, the immune system immediately mobilizes in a process called *inflammation*. As an example, the swelling, redness, and touch sensitivity we experience when we get a burn, a deep cut, or a recalcitrant splinter is a natural and necessary response to the presence of a wound and/or foreign substance. This response is evidence of the workings of the cells of the innate and adaptive immune systems, which are mobilized to the injury site to destroy microorganisms and orchestrate the repair of the wounded tissue, including the removal of dead cells and cellular debris. Furthermore, the inflammatory response is carefully regulated to ensure that after the cellular damage that engendered the response is repaired, the response is shut down so that healthy cells are not damaged. Such acute inflammation is a short-term response that rapidly restores the tissue to its normal, biologically balanced state of *homeostasis*.

As the wound successfully heals, the inflammatory response subsides. However, if an irritant is not effectively removed, and thereby persists over a lengthy period, inflammation can become a chronic condition. In a case of continuous irritation over prolonged periods—cigarette smoking, for example—the chronic inflammation at the site of the exposure—in this case, the lungs—significantly increases the likelihood of developing cancer. In the case of lung cancer, the top cause of cancer fatalities, while the damage to the DNA inflicted by smoking is no doubt involved in tumor development, the presence of chronic irritation drives inflammatory processes that promote tumor initiation and growth.

Over the years, as we accumulate increasing amounts of cellular damage from environmental irritants and toxins, our immune competence degrades, and a low level of chronic inflammation often develops in which the immune response is activated slowly over time. The immune system adjusts over time to continuous exposure to the antigen by dampening the response to that antigen. This leads to a reduction in the responsiveness and capability of the innate and adaptive arms of host immunity. Like "the boy who cried wolf," over-exposure to danger signals can result in a dulling of the danger response.

At the cellular level of the innate mechanisms of immunity, aging brings a decrease in the activity of *natural killer* (*NK*) *cells*. Like T cells, natural killer cells are capable of directly destroying microbes and tumor cells, although their mode of activation differs from that of T cells. Rather than recognizing specific foreign antigens like T cells (and B cells), NK cells recognize molecular patterns on cell surfaces characteristic of foreign and damaged cells (including virally infected cells) that might, in turn, signal the presence of a potential pathogen or tumor cell. Since NK cells do not need to "learn" to recognize a specific antigen, they are part of the innate, rather than the adaptive, immune system.

As we age, we also suffer from a reduction in the activity of *dendritic cells*, which are responsible for coordinating the communication between the innate and adaptive immune systems. During the inflammatory response, dendritic cells bind proteins in the region of an infection or wound (or tumor). The dendritic cells, which are known as *antigen-presenting cells*, or *APCs*, deliver the proteins to a membrane-bound cellular compartment called a *lysosome*, where the proteins are degraded into small pieces.

These small protein fragments (of about five to seven amino acids) are returned to the dendritic cell's surface, where they bind to proteins found on the surface of most of our cells. These proteins are called *histocompatibility locus antigens* (*HLA*). This highly diverse set of membrane proteins is the product of genes found in the *Major Histocompatibility Complex* (*MHC*). The MHC, which is found in humans on chromosome 6, contains two subtypes of genes, called Class I and Class II. The genes of the MHC are responsible for the regulation of the immune response, and they are amongst the most diverse genes inside of us, with hundreds or even thousands of sequence variants (called *alleles*) for each gene resident in the human population.

In a process called *antigen presentation*, the APC brings the antigen fragment, bound to an HLA molecule on the surface of the APC, in proximity to the T cell surface, where a complex biochemical interaction takes place. In this interaction, the T cell, by the miraculous precision of biochemistry, determines whether the small peptide fragment is derived from the body's own complement of proteins, or whether it has an exogenous origin.[29] By so doing, the immune system is trained, starting early in life, to recognize harmless "self" antigens from "non-self" antigens that may pose a threat.

The phagocytic cells of innate immunity—the neutrophils and

macrophages—which mediate the destruction and removal of foreign cells and cellular debris, also undergo a diminution in their power as we age. As the functionality of the innate immune system decays, the adaptive immune system also undergoes changes because of the degradation of its cellular components.

The development of an effective adaptive immune response requires not only cooperation with the innate immune system, but it also requires the presence of a population of healthy precursor stem cells (described in Chapter 2) that can be stimulated to form B and T cells, which are subsequently primed by exposure to antigens.

Aging brings on a reduction in the entire population of precursor cells for the B and T cell lineages such that it becomes more difficult, with age, to develop activated B and T cells specific for the vast multitude of antigens that we are exposed to every day. In addition, due to the lower activity of the dendritic cells (noted above), a gradual loss of B and T cell reactivity is inevitable.

In a healthy individual, the innate and adaptive responses work in harmony to eliminate foreign cells and thereby maintain homeostasis in our tissues. The various cells of host immunity routinely detect and destroy foreign invaders and tumor cells in a process called *immunosurveillance*, a round-the-clock "seek and destroy" mission by which our immune cells search for the presence of foreign antigens. Effective immunosurveillance is critical to our well-being.

Current data suggests that small tumors that are invisible to all our current means of detection form spontaneously inside us. In most cases, they are detected and destroyed by immunosurveillance. It is believed that such "invisible" cancers, known more formally as *covert cancers*, form with regularity. It is only in the rare cases where immunosurveillance fails that a tumor progresses from a small mass, tens of millions of cells, to a potentially malignant cancer.

Cancer cells that can avoid detection by the host immune system will naturally survive in preference to those that are subject to recognition and destruction by host immunity. In a process called *immunoselection*, vulnerable cancer cells are culled from the body by host immunity, leaving behind other cancer cells with the ability to resist immunosurveillance. Such cancer cells are hidden from the host defenses—cloaked, as it were, from the sensors of host immunity. Within the past decade, clinical evidence has emerged that demonstrates the tremendous value of the decades of intensive research into cancer immunity. This evidence can be found in the burgeoning field of *cancer immunotherapy*. For the first time in human history, we have found a way to mobilize the immune system to detect and destroy cancer cells hidden from host immunity. The excitement in the scientific and medical communities is palpable, for reasons that will become clear in subsequent chapters.

Chapter 2

The Plastic, Fantastic Cell

In the fall of 1953, Leroy Stevens was a 33-year-old postdoctoral researcher working in his first job as a professional scientist since graduating from the University of Rochester with a doctorate in embryology. A native of Kenmore, New York, a suburb of Buffalo, Stevens attended Cornell University before pursuing his Ph.D. After earning his degree, Stevens accepted an offer to join the Jackson Laboratory in Bar Harbor, Maine. The Jackson, an independent research institute, boasted the largest mouse-breeding colony in the world. A laboratory with such a precious resource was a wonderful place to start a career as an embryologist.

Embryology—the study of how the fertilized egg grows into an independent, free-living organism at birth—relies heavily on *Mus musculus*, the common house mouse. The mouse has long been a highly valued model system for human biology. *Mus musculus* is also a key animal model used in the study of *embryogenesis*, the astounding cellular ballet that takes the fertilized egg through the well-ordered steps of development, resulting in a mature and viable organism capable of living freely in nature.

The process of embryogenesis begins with the union of the gametes, the egg and the sperm, at fertilization. The fertilized egg (known as a *zygote*) then undergoes an ordered series of transformations that result in the development of a free-living organism. These transformations are driven by the genetic information passed to the zygote from its parental cells (the egg and the sperm), resulting in the formation of all the cells required for the zygote to develop into a biologically mature organism. In humans, this transformation process takes about 280 days. During that time, the developing fetus goes from a single cell to a fully developed newborn baby containing trillions of cells.

Stevens' work was funded by a tobacco company intent on discrediting a claim that was beginning to inflict damage on their booming business enterprise: the assertion that cigarette smoking causes lung cancer. The tobacco interests were aware of the significance of the growing body of data supporting the contention that smoking tobacco can have serious, and even fatal, health effects.

Despite convincing evidence from multiple sources indicating the presence of chemical compounds in tobacco known or suspected to cause molecular damage in the lung, the bigwigs in the tobacco industry nonetheless intended to demonstrate that tobacco itself is not a causative agent of cancer. Rather, their hypothesis contended the fault could be found in the byproducts of the paper wrapper's combustion. According to this line of reasoning, cancer was not caused by tobacco but rather by the delivery vehicle itself. Once the paper's toxic byproducts were identified and eliminated, cigarette smoking would be freed from the allegations about its potential link to lung cancer.

It was difficult to understand the reasoning behind the proposal. It sure seemed a stretch, at any rate. Was the problem due to a substance containing hundreds of chemicals and byproducts of combustion or, rather, was the cigarette paper the culprit?

We might call this hypothesis scientifically unconvincing, but that might be too generous considering the most distressing blemish on the proposal: there wasn't any credible scientific data to support it. This is a fatal flaw in any scientific argument. We should, however, acknowledge the creativity under duress (not to mention the audacity) required to put the proposal forward to the public.

We might even say the idea stank of desperation. However, the tobacco interests bought a little time by asking for more studies. Not much, however, as Stevens, an eager young investigator, dove right into the problem. His approach to evaluating the hypothesis was logical, straightforward, and easy to implement in the laboratory. He separated the tobacco from the paper and used a variety of solvents to extract the chemical constituents of the cigarette's components in isolation from each other. With these test articles in hand, he could apply the isolated extracts directly on the skin of laboratory mice and evaluate the biological impact on both the mice and their offspring.

In this way, the experiment could evaluate not only whether the test materials were *carcinogens* (cancer-causing agents) but also whether the materials were *teratogens*. Such substances cause genetic damage to the fetus in the form of congenital malformations, also known as birth defects.

The experimental hypothesis was easy to disprove (as the data available at the time foretold even before the first test tubes were mixed). In short order, the experiments clearly demonstrated that tobacco is a carcinogen, a substance that leads to the development of cancer in exposed animals.

While tumors were evident in many of the experimental animals, meticulous examination of thousands of descendants of the exposed mice did not demonstrate the presence of birth defects. However, he did find something far more intriguing. One day, while examining a young mouse derived from a lineage known as "Strain 129," Leroy came across something exceedingly rare in nature and never reported in a laboratory mouse.

This mouse had a *teratoma*. Generally found in the reproductive organs where the egg and the sperm are made (in the ovaries and the testes), these bizarre cancers are amongst the strangest phenomena known to biology. Teratomas are unique from any other kind of tumor. While tumors involve the transformation of only one type of cell into a tumor cell—liver cells (*hepatocytes*) or white blood cells (*leukocytes*), for example—a teratoma contains multiple kinds of cancerous cells. Teratomas are comprised of various combinations of skin cells, muscle cells, nerve cells, bone cells, fat cells, and sometimes even wads of hair and/or baby teeth, all jumbled up together in an alien cellular mishmash.

The earliest reference to human teratomas goes back to ancient Mesopotamia. A prophecy in a clay tablet of "a woman giving birth to a child with three legs" found at the site of the Chaldean Royal Library of Nineveh (in modern-day Iraq), one of the great cities of the ancient world, has been attributed by expert testimony to "the discovery of a benign teratoma."[30] According to this prophesy dating back more than 25 centuries, the appearance of this child will bring a time of prosperity. Given the rarity of teratomas, the fulfillment of this prophecy subjected the ancient Chaldeans to an exceedingly long wait for better times.

As a trained embryologist, Leroy Stevens recognized that he had stumbled upon something of great significance: a mouse strain that was a source of a rare testicular tumor. He knew he had to put aside the (already answered) question of tobacco carcinogenicity and focus on teratoma biology.

After two years of painstaking work in his laboratory examining thousands of strain 129 mice, Stevens had isolated for further study a representative population of mice bearing the teratoma. A photograph in Stevens' 1954 paper shows a six-month-old strain 129 mouse with a large, bulbous mass between its back legs and tail, as if a dark, furry ball were sewn to its rear end. The work demonstrated the presence of the teratoma in about one percent of the strain 129 male mice.[31]

While one percent may seem like a small number, teratomas are far rarer in nature. Consequently, the inbred mouse strain 129 that produced one teratoma for each one hundred baby mice was considered a rich source of teratoma samples. Evidently, Dr. Stevens had made a wise decision by choosing to work at the Jackson Lab, where obtaining hundreds of mice for experimentation was a snap.

The teratomas' biological behavior in strain 129 mice was intriguing and perplexing. The mice were not born with teratomas. The tumors developed after birth as the testes matured, from as early as 8 days to as late as 210 days, a significant percentage of the test animals' average lifespan. Unlike the mouse shown in the photograph in the 1954 paper, there were no external symptoms of the teratoma's presence in most of the afflicted mice.

One day, while examining one of his teratomas under the microscope, Stevens suddenly realized (in what we might today call an "aha" moment) that most of the cells in the sample were elongated and variable in shape and overall size. These were the developed cells expected in the mouse testes. Scattered amongst the many elongated cells, he noticed another cell type that did not share the same features as the fully developed cells. These cells—and there were only a few—were smaller and rounder and lacked the elongated appearance of most of the cells in the microscopic field.

Stevens recognized that these cells had the appearance of the cells in an early embryo that had not developed into the specialized cells characteristic of a fully developed organ. Something had gone terribly wrong during the development of the fetus. He realized that these round cells must be embryonic cells that were supposed to mature into the specialized cells of the testes. Instead of undergoing development, the embryonic cells had remained in their embryonic state and had not undergone the expected biological transformations coded in their genes.

The ramifications of this finding were astonishing. Here, in a colony of laboratory mice, Stevens found a reliable source of embryonic cells. These cells were ideal for studying mouse embryogenesis—and more. Because the embryonic cells from the teratomas could be used to investigate what goes awry during the tumor's genesis and growth, he realized that this line of investigation might open a window into the mechanism of tumor formation.

In a staid manner—consistent with reports of Leroy Stevens' understated and humble nature—the last sentence in his 1954 paper specifies the importance of further work using strain 129: "It is pointed out that an inbred strain of mice in which a relatively significant percentage of males develop testicular teratomas may be an important tool in the study of some hitherto unexplored aspects of the biology of these interesting growths."[32]

In his early experiments, Stevens found that when he minced tissue from a strain 129 teratoma, he could successfully graft the tumor into the abdomen of both male and female recipients. While the transplantation of a growing tumor was successful in all 15 attempts, the tumors in the recipients tended to grow slowly. Fortunately, 1 out of the 15 attempted transplants provided a rapidly growing tumor that could be further transplanted to new recipients, a process known as *serial transfer*. Stevens found that he could successfully transplant the fast-growing tumor, derived from a 40-day-old strain 129 mouse, every two weeks when, according to his 1954 paper, "it reaches 2 cm in diameter."[33]

The test sample in these experiments was injected into the *peritoneum*, the tissues surrounding the abdominal organs. When tumor cell transplantation is successful in this procedure (known as *intra-peritoneal transfer*), the tumor is established and grows in the peritoneal space. Within days, a firm, fluid-filled sac of tumor cells arises in the gut, as if the tumor has built a cocoon full of nutrients around itself to provide the molecular components required for sustained growth.

Turning his attention turned to the embryonic cells in the tumor, it seemed logical to hypothesize that these cells—endowed with a "superpower" that enables them to give rise to many types of cells during embryogenesis—held the key to understanding teratoma biology. Reasoning that "the preponderance of embryonic cells in these relatively simple tumors makes it seem likely that they give rise to the diverse types of differentiated cells found in the more complex teratomas," he began to develop a concept of how teratomas form.

Stevens proposed that a teratoma occurred because of a failure to properly execute the full developmental program of the embryo. Since the teratomas in strain 129 mice "are composed of a variety of embryonic and adult tissues which are not normally found in the testis," some of the embryonic cells must have malfunctioned in the execution of their genetic programs. Instead of producing a well-organized fetus with the appropriate types of cells and tissues throughout the organism, defects in embryonic cells created a disordered cellular hodgepodge in the testes of the newborn mice stricken with this rare tumor.

When he transplanted teratomas into the testes of adult mice, he found that he could reproduce the teratomas in mature animals. This astounding finding suggested the teratoma—or, at least, some component of the teratoma—can seed a new tumor in a fully developed animal. This work provided direct evidence that the aberrant embryonic cells in the teratoma sample were responsible for transmission of the tumor. He called these *embryonic carcinoma (EC) cells* in recognition of their ability to seed new tumors.

Seven decades ago, Leroy Stevens provided a conceptual framework for teratoma formation and a reproducible mouse model system for studying embryonic stem cells that enabled additional investigations into the nature of tumor development. The data supported the mechanistic link proposed by Stevens between molecular errors during embryogenesis and the formation of teratomas in young animals.

The reverberations of the contributions of Leroy Stevens that demonstrated the relationship between aberrant cell development and cancer are still strongly felt today. By developing the teratoma model, Stevens directly showed that genetic errors during cellular development can lead to cancer. Equally important, Leroy Stevens provided, for those who followed in his pioneering footsteps, the means for obtaining precious embryonic cells for laboratory experimentation.

These developments also provided the basis for systematic laboratory experimentation with stem cells from embryonic tissues, giving rise in the following decades to the burgeoning field of stem cell biology that today promises the possibility of replacing damaged cells with functional equivalents. As an example, stem cells implanted into the pancreas can create new pancreatic islet cells to produce insulin in diabetics. The possibilities seem endless.

Despite these outstanding contributions to biology and medical science, the name of Leroy Stevens has been lost to memory, even amongst many biologists. Looking back on his career from his home in Vermont in 1990, Stevens noted, in his characteristically unassuming manner, "This stuff was extremely interesting, and it sure beat studying cigarette papers!"[34]

We should all be grateful for the life and work of a man of great insight, determination, and humility, Dr. Leroy Stevens, the accidental and unsung hero of cancer biology.

From the moment of conception until death, the creation and maintenance of the distinct types of cells in the body, and the replacement of existing, damaged cells, is subject to tightly regulated mechanisms. These mechanisms are driven by pre-existing genetic programs—subroutines, in the language of computer programming—that ensure homeostasis in multicellular organisms like us.

It all starts with the fertilized egg, the zygote, a single cell containing everything required to create a fully developed organism. The zygote is a *totipotent cell* (*toti* means "whole" in Latin), a cell that can give rise to all the cell types required to develop a mature organism. In addition to being the progenitor cell of all the cells of the embryo itself, totipotent cells are also the ancestors of all the tissues outside the embryo (*extra-embryonic tissues*) that support the embryo's growth, such as the placenta.

Once the zygote starts to divide, it creates *embryonic stem cells* (*ESCs*), the precursors of all the cell types of the fully grown embryo. The ESCs are born within a few days after embryo formation (beginning about four days post-fertilization in humans). The path of stem cells from their birth in an embryonic structure called the *blastocyst* to the formation of the various cell types found in the body occurs in discrete steps.

ESCs do not directly give rise to specialized, tissue-specific cells in a single cellular transformation. Rather, an ordered, multi-step process takes place, whereby the ESC creates additional sets of stem cells that are the direct parents of the specialized cells found in the body's mature tissues. An illustrative example of this process (shown in Figure 3) depicts *hematopoiesis*, the differentiation pathways for the cells of the blood and lymph that takes place in the bone marrow.

Hematopoiesis

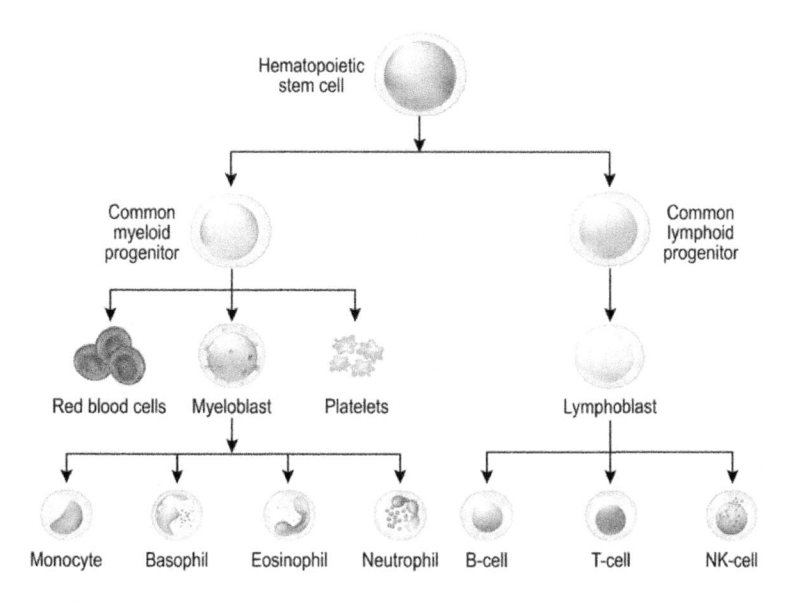

Figure 3. Schematic diagram of hematopoiesis

The *hematopoietic stem cell* (*HSC*)—itself derived from an embryonic stem cell—is the precursor cell for all the cells of the blood and the lymph, the fluid which flows through the lymphatic system that serves as a distribution network for immune cells. All these cells are derived from stem cells called the *common myeloid* and *common lymphoid progenitor cells*. The former is the parent of red blood cells, platelets, and other myeloid cells that participate in innate immunity. The lymphoid progenitor is the parent stem cell of B cells, T cells, and natural killer (NK) cells.

The ESC is known to cell biologists as a *pluripotent cell*, denoting its potential to give rise to stem cells capable of generating all the cell types in the body but not the extra-embryonic tissues (*pluri* means "many" in Latin). The offspring of these pluripotent cells, such as the hematopoietic stem cell shown in Figure 3, are also stem cells. These cells have more limited potential than pluripotent cells because they generate a limited set of cells rather than all the cell types in the body. These are known as *multipotent cells*.

Multipotent stem cells can generate specialized cells with defined functional roles. These are the workhorse cells that perform our bodily functions. For example, the hepatocyte (liver cell) can support the functions of the liver, such as the chemical inactivation of toxins. The myocyte (muscle cell) is endowed with the ability to contract on demand in response to electrochemical stimulation by the nervous system.

The specialized differentiated cells are produced from their stem cell parents in the process of *differentiation*. These specialized cells have distinguished, or differentiated, themselves functionally from their parental cells. The resulting differentiated cells produce offspring that are copies of themselves; they are *unipotent*, with the potential to generate only one type of cellular offspring, a copy of itself.

The ESCs' genetic program is activated only during the embryo's development, transforming all the ESCs into the various tissue-specific, pluripotent stem cells. The tissue-specific stem cells serve as long-term reservoirs that can produce new, differentiated cells when they are needed throughout the life cycle. In this manner, the body can rapidly build new cells to repair a wound and/or replace damaged cells. In addition, because differentiated cells can undergo only a limited number of cell divisions before they require replacement, the hierarchical network of stem cells can provide new sets of differentiated cells in any tissue of the body that needs them.

The biological processes that govern the cell's path from birth to death determine its fate. The rate at which the cell reproduces, and the progressive steps that bring it from the original embryonic cell to full maturity in the differentiation process, are governed by genetic control mechanisms that determine which proteins are manufactured by the cellular machinery. For any cell, control mechanisms ensure that the right amounts of the various proteins for the given tissue are available at the right times and, most critically, in the right places of the body. "Beginning from a single totipotent cell, successive waves of self-renewal, differentiation, and commitment ultimately yield the intricate array of cell types, tissues, and organs of a fully formed organism."[35]

Stem cells have another unique property that distinguishes them from other types of cells. Unlike differentiated cells, stem cells can divide for years or decades, perhaps even throughout the entire lifespan. Biologists call this property the *capacity for self-renewal*. Coupled with the ability to produce more than one type of progeny, the capacity for self-renewal provides a functional definition of a stem cell.

The magical cell uses a finely orchestrated set of biochemical mechanisms to choose which genes are activated (and, equally important, which are turned off) at any given time and place. Such control over what biologists call the expression of the genes (aka *gene expression*) lasts throughout the cell's life cycle. From the same set of 21,000 or so protein-expressing genes, the human cell can alter its gene expression to render the phenotypic properties of about 200 different cell types, each with its own unique patterns of gene expression and protein production. These control mechanisms are responsible for the outcome of embryogenesis and the maintenance of the appropriate cell types in the various tissues throughout the lifespan.

The concept that cellular programming can be altered to change a cell's phenotype (appearance and functional properties) is called *cellular plasticity*. This does not mean cells are made of plastic; in this context, plasticity refers to the "capacity for being molded or altered."[36] In biological terms, cellular plasticity indicates a (generally reversible) ability to change appearance (morphology) and functional characteristics, as determined by differences in protein expression that provide for differences in size, shape, and physiological function.

All the genetic programming to provide for the activities of every cell type in the body is encoded in the DNA upon conception. Each cell type uses only a portion of the total genomic information present in the embryo's DNA. For each cell, its identity and function are defined by its patterns of gene expression over time.

So how does the cell perform these amazing feats of individuality, selecting the appropriate programs from the unfathomably complex genetic menu that encompasses the instructions for every cell type, under all conceivable conditions, throughout the entire lifespan?

Had you asked this question in, say, 1952, when Leroy Stevens was actively pursuing the secrets of the teratoma, the answer would have been based on pure speculation. However, in the next year, the door to understanding the genetic secrets of the magical cell was cracked open in the Cavendish Laboratory at Cambridge University.

In April 1953, a short publication by James Watson and Francis Crick in the British journal *Nature* proposed the double-helical structure of DNA.[37] This paper, which contained little data and only a single diagram of the DNA double helix drawn in pencil by Crick's wife, Odile, changed biological research and the practice of medicine forever.

Chapter 3

The Caretaker and the Gatekeeper

Evolution has provided an extensive array of mechanisms that allow our cells to accomplish the biochemical activities that sustain life: absorbing nutrients, excreting waste, controlling cellular metabolism, creating the building blocks of biomolecules, and monitoring and repairing (or, if necessary, destroying) damaged cells.

To execute these processes, biological systems rely upon an elaborate communications network to send and receive messages within the cell, a process known as *intra-cellular communication*. In addition, cells communicate with each other in a process called *inter-cellular communication*. These communication networks operate at dazzling speeds, such that the cellular circuitry can rapidly modulate biological outcomes to promptly respond to changes in the cellular environment. The ability to rapidly respond to changing conditions is essential for survival. Whether the stimulation is a minor temperature change, which may render only a mild biological impact, or an unexpected danger requiring an instantaneous "fight or flight" response, the cellular communications networks have evolved to remain continuously vigilant, much like the operational systems at a nuclear power plant.

While the plant itself cannot alter the operational status of a nuclear reactor—thereby requiring human intervention or, at the very least, automated electronic systems built and programmed by their creators—biological systems require no external intervention or instructions to assist them in responding to environmental changes. They are evolutionarily equipped to control this critical function around the clock. There is no time for shutting down the system for repairs and no breaks for the molecular "workers" throughout the lifespan.

We can only marvel at such natural wonders. Using tiny amounts of electrochemical energy, biological systems can orchestrate all their myriad functions on a continuous basis, adapting as required to environmental stimuli. Moreover, biological systems accomplish these functions with fantastic efficiency. As a prime example, the human brain, the center of our consciousness and the biological nerve center of the body, is powered by electrochemical circuitry operating in the 100-millivolt range (or less). This is one-tenth of a volt—less than 10% of the output of a single one-and-a-half-volt AA battery.

Despite operating at this low energy demand, an insignificant fraction of that used by our computers, the return of the enterprise is beyond comprehension: brilliant scientific theories; unforgettable art and literature; musical masterpieces; the ability to think, feel, and love. Such efficiency in performance, so much "bang for the buck," is far beyond our current ability to replicate using even our most advanced electronic technologies.

We live (for better or worse) enveloped in a world of electronic information. In our "Brave New World" of high-speed information technology (IT), the computers and servers that support the network (the computer hardware) function as central repositories of knowledge, the brains of the system. There, the data is stored, in perpetuity, ready for retrieval when so instructed by the software. The software contains an array of instructions that regulate the operation of the system.

To function effectively and with low error rates, the computer hardware and software must be seamlessly integrated and thoroughly evaluated before deployment to ensure high performance over extended periods and under changing conditions. As we all know from experience, these efforts usually bear fruit, but not always. Bugs that impact system performance and lead to a crash are always part of the picture, events that impact the system's functionality. In addition, the software is vulnerable to infection by computer viruses, malware, and other modes of malicious mischief, now ubiquitous in the IT environment, which threatens the operational integrity of the system.

The machines do not make calculation errors. They simply follow the programming as written, actuating the instructions in the order dictated by the software. Occasionally, the programming does not provide clear instructions for the execution of succeeding steps, perhaps because the operating software faces a situation it has not seen before in development or during the debugging of the software. Under such conditions, the operating system can encounter a conflict in the instructions that precludes a clear operational decision.

The computer hardware is a product of its creators, so the error on the screen can only be ascribed to one source: not the machine, but rather, the software itself. In our greatest moment of pique, as we are poised to begin swearing at the device, we may forget that the computer cannot proactively correct the mistakes of its human creators. With apologies to Shakespeare, the fault dear user is not in the stars, or in the machines, but rather, in ourselves.

At the level of fundamental physics, the communication of information through a computer system results from the transfer of electrons. The flow of electrons through the circuitry is how messages are relayed from point to point in a simple language that uses a binary code. Each decision, each calculation, provides one of two results: zero or one yes or no—with respect to the decision of whether electrons will travel down one path or another. These decisions are managed by digital switches in the microchips that divert electron flow along the desired path(s) until the electrons reach their destination(s) and trigger the desired actions directed by the software code.

The biological world also has a language of its own. The language of biology is spoken by molecules. To fully appreciate the molecular scale of proteins—the large molecules synthesized by our cellular machinery to conduct the cell's molecular business—we can start by envisioning a simple molecule like acetylsalicylic acid. You know it as aspirin. The aspirin molecule contains three types of atoms—carbon, hydrogen, and oxygen—and 21 atoms in total, with a *molecular mass* of 180.159.

The molecular mass reflects the total mass of the protons and neutrons in the nuclei of the atoms that comprise the molecule. The units of molecular mass are called *Daltons* in recognition of the contributions of British chemist (and schoolteacher) John Dalton to the development of atomic theory. One Dalton is defined as one-twelfth the mass of a carbon atom—close to the mass of a hydrogen atom, which is comprised of one proton and one electron.

Our proteins contain 20 types of *amino acids* with an average mass of about 110 Daltons. Amino acids are comprised of up to six types of atoms, rather than the three types found in the aspirin molecule.[38] A protein may contain hundreds, or even thousands, of amino acids in its sequence, with molecular masses ranging from the thousands to more than one million Daltons.[39]

Proteins come in a vast array of shapes and sizes, with the two major categories broken into *fibrous proteins*, which have elongated structures, like pieces of rope, and *globular proteins*, which have more spherical structures. Examples of fibrous proteins include actin and myosin, the major protein components of muscle fibers, as well as a family of proteins known as *extracellular matrix* (*ECM*) *proteins*, which form a structural net for the cell known as the *cytoskeleton*. These proteins, which provide for the cell's structural integrity, are also essential for cellular motility, the movement of a cell across a surface.

Many globular proteins are *enzymes*. These biochemical catalysts lower the energy requirement for chemical reactions inside the body, vastly increasing the speed of these reactions. A chemical reaction can occur if the participants in the reaction, the *reactants*, wander aimlessly and just happen to find each other. This would be a random event that would occur infrequently, and the reaction would occur very slowly. The job of enzymes is to actively seek the reactants and bring them together in a spatial configuration relative to each other that lowers the energy required for a chemical reaction to take place, thereby increasing the speed of the reaction. In this process, called *enzymatic catalysis*, an enzyme binds to its reactants and then changes its three-dimensional shape to bring them together. Enzymes are essential to life; without them, biochemical reactions would be too slow to support the rapid and dynamic processes that take place in living systems.

Also amongst the globular proteins are molecules that bind to specific molecular targets throughout the body. Many are signaling proteins that bind to other proteins inside the cell or on cell surfaces, where they bind to surface proteins called *receptors* to elicit a biological effect inside the cell. Hormones such as insulin fall into this class.

Antibodies are also globular proteins that bind molecular targets in the circulation or on cell surfaces. These large proteins bind tightly to their targets, inhibiting the target's function. In the case of antibodies that bind to proteins on the cell surface, the result of antibody binding can be the destruction of the cell bearing the target molecule. This highly specific immune response protects us from disease-causing (pathogenic) bacteria, fungi, protozoans, and other potential invaders, including viruses.

Despite their complexity, the underlying structural concept of proteins is simple: a chain of amino acids, strung together by chemical bonds called *peptide bonds*, with one of the 20 types of amino acids (in humans) at each position of the chain (Figure 4).

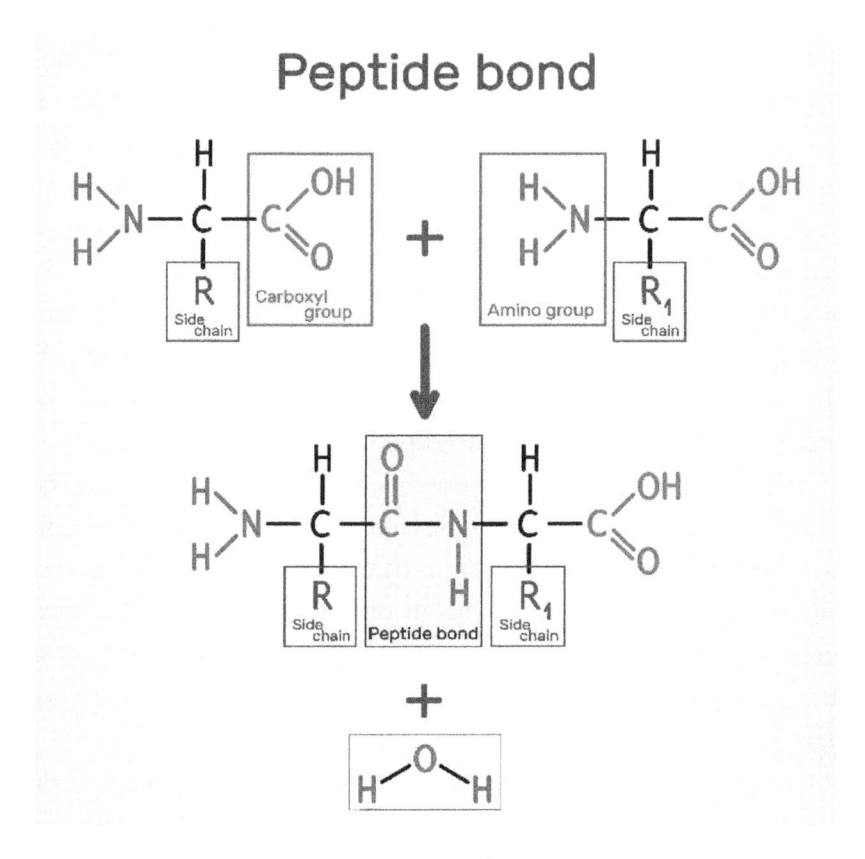

Figure 4. Formation of the peptide bond

Peptide bonds (bottom, center) form between a carbon (C) atom on one amino acid and a nitrogen (N) atom on another amino acid (shown in the center of the bottom figure), with the release of water following the formation of the bond. Amino acids (there are two at the top) are comprised of an amino group linked through a central carbon atom to a carboxyl group. The differences between the amino acids are based on the chemical nature of the "R" group that projects off the central carbon atom in the structure. These R groups, known as side chains (thank you, Paul Ehrlich), vary in structure from a tiny hydrogen atom (this is the amino acid glycine) to the more complex and far larger molecular structures of amino acids such as tryptophan and phenylalanine (and no, there won't be a test).

In chemical terms, a peptide bond is a type of *covalent bond*, in which two atoms share electrons.[40] Covalent bonds are stronger than *hydrogen bonds*, in which a positively charged proton interacts with another atom (oxygen or nitrogen in biological systems).[41] Covalent bonds are orders of magnitude stronger than ionic bonds, such as those between sodium and chloride atoms in table salt. Ionic bonds are so weak that the atoms are no longer bound together upon dissolution in water.

Carbon-nitrogen bonds are particularly hardy. To break the peptide bonds in a protein in the laboratory, one must heat the sample above the boiling point of water in high concentrations of acid or base.[42] Peptide bonds are the glue that holds our proteins together. Without the ability of peptide bonds to maintain their structural integrity while performing their cellular work, it would be impossible to sustain life. While proteins vary significantly amongst the myriad forms of life that populate our planet, the basic principles of protein structure and function are ubiquitous in nature.

Thus, from this simple set of ingredients, living systems can potentially create a limitless array of proteins, each with a unique order of amino acid building blocks that have been "built-for-purpose" over evolutionary spans of time to fulfill their biological functions.[43]

Thanks to the discovery of the DNA structure, we know that the genetic material is a twisted ladder in the form of a double helix that contains the genetic code, which is comprised of four different chemical units, collectively called *nucleotides*, and known by the letters A, C, G, and T. These letters stand for the *nucleotide bases* adenine, cytosine, guanine, and thymine, respectively. The amino acid sequence of a protein is dictated by the sequence of nucleotide bases on the DNA that together comprise the genetic code.

Along the double helix, an A on one strand always pairs with a T on the adjacent strand, and G always pairs with C. While these *base-pairing rules* are simple, the nucleotide base strings are rich in content. In these nucleotide base arrays, we find the entirety of the biological programming that takes us from the zygote to our very last breath.

The linear sequence of nucleotide bases along the strands of the double helix contains all the information required to produce all the protein molecules that comprise us, correctly folded in complex three-dimensional arrangements governed by the laws of chemistry. It is like having a piece of paper that provides the instructions for its own folding into a complicated Origami structure, all accomplished without the need for someone to do the folding.

While we know the goal of protein folding—to attain the three-dimensional orientation where the protein expends little or no energy to maintain its structure and the shape in which it can maintain its full biological activity—we cannot predict the three-dimensional folding of the protein based on the sequence information alone.[44] Even the most sophisticated computer models can provide only approximations for predicting protein structures in three-dimensional space. Notwithstanding our modern analytical tools, and the power of our computers, proteins remain mysterious in terms of their structural dynamics, an endless source of fascination for the biophysicists and structural biochemists who study them.[45]

Inside the cell, in the parlance of electronic communication, we can consider the DNA as the central computer, the cell's central server. Inside that server, we find, in the DNA sequences, "software" that encompasses two types of information: that required for forming the full-grown organism from the embryo and that needed to support the biological functions of the organism throughout its lifespan. The biological software is responsible for ensuring the programs that orchestrate cellular events are implemented in the right place, at the right time, in the right order, and for the proper duration. It contains the totality of the information required for properly executing the genetic program over the cell's entire lifespan. And unlike the computer, which requires human intervention when its systems go amok, no outside intercession is needed for the cell to maintain and repair the biological operating systems encoded in the DNA.

As stated previously, the functionality of computer systems is based on a binary code—one or zero, on or off, yin or yang, yes or no. The biological world, in contrast, utilizes far more sophisticated coding algorithms to transfer information. The genetic code is constructed using the four nucleotide bases, represented by the letters A, C, G, and T—with one of the four nucleotide bases at each position of the coding sequence. By joining these four bases in various combinations, the code produces a far more information-rich output than is possible with a binary code of zeros and ones. And from these four simple biochemical ingredients, the entire biological world springs forth in all its glorious diversity and splendor.

———

The molecular signaling required to operate the cell, a process known to biologists as *signal transduction*, is achieved through intricate networks of molecular circuits that intertwine and intersect to provide multiple (and sometimes overlapping) routes to the required biological outcomes. This design increases reliability and provides system redundancy in the event of a "short circuit" in a pathway due to damage or disease.

The proteins, nucleic acids (DNA and RNA), carbohydrates, and lipids (fats) that work together to accomplish the biochemical work in the cell—collectively known as *biomolecules* or *macromolecules*—interact in highly specific ways, in every conceivable combination, to control biological signaling. These interactions between various combinations of proteins, DNA, RNA, carbohydrates, and lipids provide exquisite biological discrimination, what biologists call *specificity*. Due to this specificity, each macromolecule interacts with the right partner molecule(s), thereby ensuring the reliability and reproducibility of the molecular interactions in living organisms.

These precise macromolecular interactions are dependent not only on the biochemical compositions of the macromolecules but also on their shapes in three-dimensional space—what biochemists call their *conformations*, or, in more formal biophysical terms, their *higher-order structures*.

Biological specificity is possible because biomolecular interactions occur in three-dimensional space rather than the simpler case of a two-dimensional interaction between a pair of flat surfaces.

The spatial requirements required for fitting together a pair of three-dimensional shapes are far more demanding than those for a pair of two-dimensional objects. Think of the difficulty of assembling a 2000-piece jigsaw puzzle; next, imagine the puzzle pieces are three-dimensional. That would be a far more difficult puzzle than one assembled in two-dimensional "Flatland."

The operating capability of our computers can be damaged by a limited number of known agents: magnetism, moisture, heat, and particulate matter (dust), for example. If we protect our computers from such conditions, we can expect them to operate properly (at least in the absence of software defects). In contrast, the software in biological systems—the information encoded in the genome—is under constant assault by environmental stresses that can lead to permanent changes in the programming over long spans of time. Our DNA, and that of all species, is designed to respond to the challenges of the environment and to change in response over time.

Conversely, a computer will not adapt to a damaged pathway by generating a new one simply because it cannot do so. It cannot fix its own bugs, which are invariably built in (accidentally) by the designers. Because of the malleability of our programming, however, we, like all biological organisms, are capable of remarkable adaptation. We can repair (at least some of) our defects—the healing of a wound, for example.

We can even, on occasion, correct our own mistakes. If an error occurs during DNA replication, it can usually be repaired. However, not all errors are successfully repaired, nor is such perfection possible in living organisms. Without errors that lead to changes in DNA sequences that are passed on to offspring, there would be insufficient genetic variation in populations to drive evolution. On occasion, sufficient errors will lead to cancer, such that cancer is, unfortunately, a natural and unavoidable risk of species evolution.

Given the genome's critical biological role and the need of every cell to access, at any given time, only a subset of the genome, it was evident (even in the early days of molecular biology) that there had to be inherent biochemical processes orchestrating gene expression. Six decades hence, we now recognize that gene expression is regulated by an elaborate set of intertwined molecular processes responsible for the execution of every genetic program accessed throughout the lifespan.

The pioneers of molecular biology could not have predicted from Watson and Crick's simple DNA model that the molecule in real life is extraordinarily fragile. Our genetic material is in a continuous state of unremitting peril, from the sun's relentless ultraviolet radiation to the pervasive contamination of our air and water with chemical toxins. The cellular quality control systems charged with managing the DNA's biochemical integrity (described below) face persistent challenges in orchestrating the critical task of detecting and repairing damage to the genetic material.

Coping with a constant barrage of environmental insults is not a new requirement for the survival of terrestrial species such as *Homo sapiens*. Damage from the sun's ultraviolet light has been a source of environmental stress throughout our evolutionary history. As a result, efficient biochemical mechanisms have evolved to assess and repair DNA damage induced by exposure to solar radiation.

Not all the damage results from external factors, however. As a result of our metabolism, byproducts are produced that can damage our cells' molecular constituents. Ironically, one of the most potent threats to our biological molecules results from byproducts of biochemical reactions involving oxygen, an essential ingredient for all *aerobic* (air-breathing) life on Earth.

A stable covalent bond, such as that between each of the two hydrogen atoms and the single oxygen atom in water, is formed by the sharing of a pair of electrons by the atoms involved in the chemical bond. Shared electrons result in stable covalent bonds; unpaired electrons seek a partner and are therefore highly unstable.

The energy for cellular processes is generated by the transfer of electrons between molecules. These electron transfer events can add an electron to molecular oxygen, converting it from a stable molecule with a single covalent bond (the solid line between the oxygen atoms in Figure 5 below) to one with an unpaired electron. The conversion of molecular oxygen by the addition of an electron forms a negatively charged molecule called *superoxide* (O_2^-). Superoxide is a highly reactive molecule (called a *free radical*) that contains an unstable, errant electron (on the oxygen atom at the right of Figure 5). Since superoxide is produced from oxygen, it is a type of *reactive oxygen species* (*ROS*).

Figure 5. The superoxide radical

The highly reactive superoxide radical can readily transfer the unstable electron to a nearby stable molecule, thereby creating a new free radical (e.g., to another oxygen molecule to create more superoxide or to an oxygen atom in an amino acid on a protein). The transfer of reactive electrons from one molecule to another creates a continuing cascade—a chain reaction—that generates additional free radicals. These free radicals can both damage our macromolecules and increase inflammation in surrounding tissues.

In addition to superoxide, other reactive oxygen species are generated by our metabolism, along with multiple types of *reactive nitrogen species* (*RNS*) that pose an additional threat to our biological molecules, as the nitrogen atom is also vulnerable to free radical formation.

Evolution has provided biochemical mechanisms that allow us to cope with the onslaught of highly reactive molecules. These mechanisms rely on the ability of other molecules, called *antioxidants*, to capture the reactive electrons before they can interact with our biological molecules. For example, our cells synthesize antioxidant compounds that can scavenge free radicals, and we derive other antioxidants from nature as part of our dietary intake. Antioxidant compounds that protect us from the ravages of free radicals include Vitamins C and E, beta-carotene, minerals (including selenium and copper), and the amino acids cysteine and methionine.[46] Another important antioxidant called glutathione is a critical scavenger of free radicals. Finally, the enzyme superoxide dismutase (SOD) converts superoxide to molecular oxygen and hydrogen peroxide.[47]

Free radical damage to the DNA can cause mutations that alter the DNA sequence, thereby corrupting the genetic code. In the case of proteins, chemical changes to the amino acids caused by free radicals may decrease a protein's bioactivity by, for example, reducing the ability of a binding protein to stick to its target. Enzymes can undergo structural alterations due to free radical damage that may reduce the reaction rate of the altered enzyme. It is also possible that a chemical change to a protein may not impact its function at all.

Two major factors determine the impact of a free radical's errant electron on a protein: (1) the location in the sequence where the electron interacts with the polypeptide chain; and (2) whether the chemical change caused by the errant electron damages an amino acid critical to the protein's function and/or results in a change of the protein's conformation.[48]

Without experimental evidence, it is challenging to predict what functional effect, if any, free radical chemistry will have on the DNA. DNA damage can potentially impact DNA replication and/or gene expression because damage to the DNA can block access of the DNA replication and/or gene expression machinery to the genome.

To envision what is happening at the level of our DNA, think of the world's busiest airports, where the critical task of the pilots and air traffic controllers is to ensure the collective safety of all the airplane passengers and crew. The only way to achieve a consistently effective and safe transportation system is by following specified procedures for data sharing, communications, and collaboration, all for the benefit of air travelers. For a successful outcome, each airplane must execute its maneuvers at the right place, at the right time, and for the right duration. There is little room for error. These procedures, which are essential for successful flight operations, provide a critical quality control function for air travel.

Now imagine a runway that, like DNA, is continuously damaged. In the case of this airport runway, the damage is caused not by radiation or reactive molecules, but rather by rocks and boulders dropping from the sky. Because the damage is continuous, the repair crews must work as fast as possible, prioritizing those places on the runway most likely to see traffic first.

While the crews can keep up with the damage most of the time, on occasion, they cannot complete a repair before the runway is needed for transit by an aircraft. We might conclude that the repair process is highly reliable, though not error-free. When the damage is not repairable before an airplane lands or starts to take off, the aircraft might proceed safely, or it might go out of control and crash. There is always risk inherent in the system.

The same is true inside the cell's nucleus, though the situation is far more complicated than in the DNA airport analogy. Unlike actual runways, which are stationary and lie flat on the ground, the DNA strands are not immobile, nor are they flattened into two-dimensional space. We can imagine them as structures that twist and gyrate in three-dimensional space. At an actual airport, air traffic control can plan the timing of the relative positions of the aircraft on or approaching the field to provide appropriate safety margins between events. There are no such constraints at the DNA airport. It operates continuously, not on several runways, but many thousands at a time, with no safety margin between "touchdowns" and "liftoffs" of the proteins and RNAs that land on its "tarmac."

In addition, airports in the real world do not attempt to land aircraft right next to each other or, heaven forbid, right on top of each other. At the

DNA airport, the macromolecules move around in groups, interacting with each other and the DNA with no with no evident restriction on their proximity. The DNA airport is a bustling place.

DNA damage can have significant and even catastrophic effects on its quality. And the task of maintaining high-quality DNA is monumental. It is estimated that exposure to the ultraviolet energy in sunlight alone accounts for about 10,000–100,000 damaging hits on the DNA per cell each day.[49] A hit is defined as a biochemical change to the DNA, ranging from the addition of an atom or small chemical group to a nucleotide base to the substitution of one base for another (a *point mutation*), to the deletion of a DNA base or bases, and even to the breakage of one or both DNA strand or, more catastrophically, of both strands, the most threatening type of DNA damage.

Fortunately, we evolved under conditions that selected for the ability to repair DNA (and proteins) damaged by environmental stress. We have an extensive protein-based system of macromolecular quality controls. It seeks out damaged sites on biological molecules and either repairs the damage or biochemically tags the unrepairable molecule for degradation and recycling into new macromolecular building blocks.[50]

Damage to DNA elicits a biological response called the *DNA Damage Response (DDR)*. The DDR network is comprised of proteins that work together on an essential task for survival: locating, assessing, and attempting to repair damaged DNA.

Proteins that protect DNA integrity are known as *caretaker proteins*. If these caretaker proteins cannot successfully repair the DNA, the cell is left with a "life or death" decision. Just like the airport management needs to decide whether a damaged runway can remain in operation, the cell must determine whether its DNA is of sufficient quality to proceed into DNA replication. If not, the cellular quality control system ensures that the cell is taken out of service, eliminated via a process called *apoptosis* (also known as *programmed cell death*), a critical cellular self-destruct mechanism.[51]

Evolution is at work here. When our DNA is changed, the damage can lead to trouble. On rare occasions, a change spawns tumor formation, the process known as *carcinogenesis* or *tumorigenesis*. On rare occasions, a change renders the organism more suitable for its environment. In such a case, the individual is blessed with a survival benefit that can be passed on to future generations.

Given the ubiquitous nature of DNA damage, and the biological energy required to recycle the cell's contents and build a new cell, it is not feasible to simply trash every cell with even a small amount of DNA damage. Often, a single base change in the DNA will not result in an observable deleterious effect. Proteins are resilient to a certain amount of variability at the amino acid level. There must be some tolerance of error, or the system will crash due to the inability to maintain sufficient numbers of healthy cells for continued survival.

DNA variability is a necessary ingredient of the living world. For species evolution to proceed, DNA changes must occur. Without such changes, which are products of the stresses inherent in the environment, what biologists call *selective pressure,* evolution would come to a screeching halt.

When we consider how vulnerable our DNA is to damage, it seems miraculous that most of the time, cellular replication and gene expression operate without incident. Given the presence of tens of trillions of cells, amongst which billions replicate each day, it is truly remarkable that the cellular quality control system can ensure the integrity of the genome so effectively.

Each day, billions of cells face the critical decision of whether they should replicate their DNA and divide, creating a new cell identical to its parent in the process of *mitosis*, or, alternatively, whether they must self-destruct for the benefit of the organism. This act of self-sacrifice for the common good is something that cancer cells disregard in their relentless quest to proliferate and survive at any cost.

Biologists utilize a conceptual model to describe the molecular governance of cell growth and replication. Known as the *cell cycle*, this model of cellular behavior defines four phases organized around two key events: the replication of the DNA and the process of mitosis, which results in the creation of two daughter cells from the original parent.

The Cell Cycle

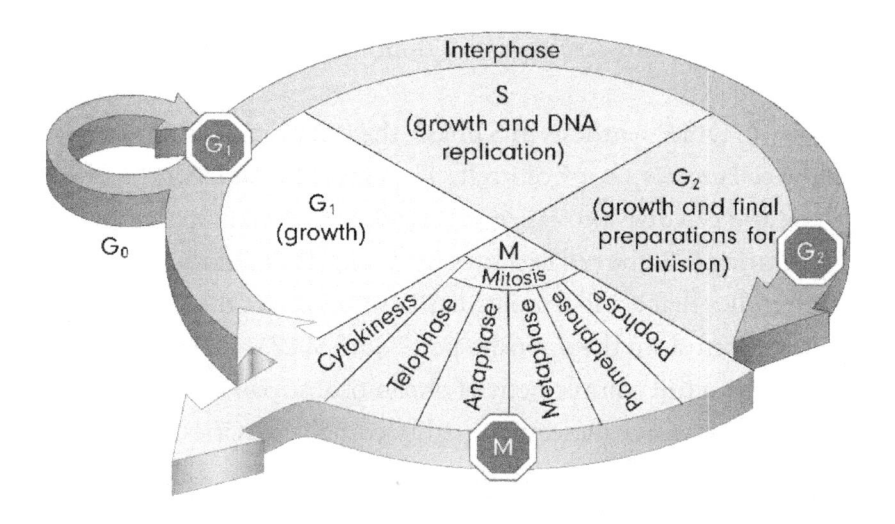

Figure 6. Phases of the cell cycle

The cell cycle's four phases are G1, S, G2, and M (Figure 6). The G1 phase, which encompasses the period between the previous cell division and replication of the DNA, is a period of internal growth for the cell. G1 (for "gap 1") is a period of high metabolic activity during which the cell manufactures the nucleotide bases and amino acids needed for DNA replication and protein synthesis. The cell is also busy during G1 synthesizing the lipids found in the membranes surrounding both the cell and some of the intracellular structures.

Examples of these organized intracellular structures, called *organelles*, include the *mitochondria*, which generate most of the cell's energy, and the *endoplasmic reticulum,* where newly synthesized proteins are processed into their mature biochemical forms.

G1 is followed by S phase, marking the point at which the cell commits to the DNA replication process. The S stands for "synthesis"—specifically, the synthesis of a new copy of the genome. During S phase, the double helix unwinds, and an enzyme called *DNA polymerase* creates a DNA strand that is complementary to each of the individual parent strands according to the base-pairing rules. This results in two double helices where once there was one.

Following DNA synthesis in S phase, the cell prepares for mitosis, in which the cell creates a copy of itself. The gap between DNA replication in S phase and cell division, called G2, is when the cell prepares for this process. During G2, the cell executes the biosynthesis of the spindle fibers, protein bundles that act like molecular lassos to direct the chromosome pairs during mitosis to the opposite poles of the cell such that each daughter cell receives a full complement of chromosome pairs. Other sets of proteins perform extensive molecular quality control activities to determine if the DNA integrity is sufficient to warrant the cell's passage into mitosis.

The ensuing mitotic cycle, M phase, is divided into the five phases that biology students (by necessity) commit to memory (at least for the test) in high school: interphase, prophase, metaphase, anaphase, and finally, telophase, during which the spindle fibers pull the chromosomes to the opposing poles of the cell. The last step of cell division, separation of the cell into two distinct new cells, is called *cytokinesis* (literally, "cell movement").

The timing of the cell cycle does not follow a regular and predictable pattern. The time between mitoses, known as interphase, can vary due to changes in local conditions. The process does not operate like a clock; there is no fixed periodicity to the progression from phase to phase. Rather, the progression through the cycle is highly responsive to the biochemical conditions surrounding the cell.

Suppose cells are challenged with conditions known to impact their biochemical processes (such as changes in temperature or oxygen concentration). In this instance, the timing of the cell cycle will be altered from that observed under homeostatic conditions. For example, as oxygen levels fall, progression through the cycle stalls, and the cell enters a non-dividing state called *quiescence (*also known as G0*)*. This state is reversible if conditions improve—or, if not, the cell is destined to undergo apoptosis.

The 2001 Nobel Prize in Physiology or Medicine recognized the importance of understanding the mechanisms responsible for control of the cell cycle. That year, the prize was awarded to the three scientists who unraveled the molecular secrets of cell cycle control: American cell biologist Leland H. Hartwell and two British scientists, Paul Nurse, and Tim Hunt. The critical processes that regulate the cell cycle are the handiwork of two types of proteins, the *cyclins* and the *cyclin-dependent kinases*. These proteins form an intricate biological signaling network that creates molecular switches that determine the control decisions throughout the cycle.

The cyclin/cyclin-dependent kinase system illustrates the simple yet flexible and responsive nature of biological control circuits. A defined sequence of biochemical events, all interdependent, ensures an action cannot proceed without molecular confirmation that the previous action has been completed. This organizational design of intertwined molecular switches provides reliability and prevents randomness in the cycle. Most importantly, it ensures the cycle meets its primary goal: to ensure that only those cells with a healthy genome are replicated to produce the next cellular generation.

The true miracle of our DNA can be found in a dichotomy. The integrity of the source code is required for the smooth operation of our cells, yet the DNA itself is fragile and vulnerable to damage, requiring a vast macromolecular network to ensure its integrity. Since the proteins in the DDR ensure the integrity of the DNA, they serve a caretaker function by providing the tools to detect and correct errors in the source code. The proteins that regulate the cell cycle, such as the cyclins and cyclin-dependent kinases, provide quality control checkpoints that ensure that the appropriate conditions are met before the process can progress to the next stage. The proteins that regulate these functions serve as *gatekeepers* for the various stages of the cell cycle.

In healthy cells, the caretaker and gatekeeper functions work together to ensure the integrity of the DNA and the appropriate progression of the cell cycle in response to environmental conditions. For example, without sufficient nutrients or other requirements for growth and DNA replication, a cell in G1 will enter G0. If conditions improve, the cell signaling apparatus triggers gene expression that returns the cell into the G1 stage, on its way toward S phase and the replication of the DNA.

In 1969, Alfred Knudson was a young pediatrician at the University of the Texas Graduate School of Biomedical Sciences in Houston when he developed an interest in childhood cancers.

Knudson was particularly fascinated by a rare cancer of the eye called *retinoblastoma*, which has an occurrence rate of about five cases per 100,000 children. While the idea that cancer might be a heritable disease held little sway at the time, it was clear that some cases of retinoblastoma ran in families, indicating a genetic factor was likely at play.

Two types of retinoblastoma have been identified. First, the familial form, which accounts for about 40% of retinoblastoma cases, has an occurrence rate of about two cases per 100,000 children. This form usually appears in toddlers; the mean age is about 15 months. It is a staggeringly malicious ailment that often involves a tumor (or multiple tumors) in both eyes.

The second form, which occurs spontaneously in the population, accounts for about 60% of cases, at a rate of about three cases per 100,000 children. In these cases, there is no evident genetic causation as determined by family history. These sporadic cases usually strike at a later stage of childhood than the familial form and, intriguingly, involve only one eye (except in the rarest cases).

Knudson wanted to know what was going on. Siddhartha Mukherjee noted in *The Emperor of All Maladies* that Knudson "wondered whether he could discern a subtle difference in the development of cancer between the sporadic and inherited versions using mathematical analysis."[52]

A budding physicist during his undergraduate years, Knudson decided to pursue a career in human medicine. Like Francis Crick, the native-born Californian was at first interested in physics and later drawn to the study of biology.[53] As a trained physicist, Knudson was able to leverage his extensive mathematical background in the effort to investigate this fascinating biological phenomenon.

Looking through the hospital records at M.D. Anderson, Knudson found 48 cases of the rare eye cancer during a 25-year period, from 1944-1969. He evaluated the data based on parameters such as patient age at diagnosis, sex, family history, whether retinoblastoma was present in only one eye or bilaterally, and, finally, by the number of discernible tumors in each eye.

Knudson performed probability calculations for outcomes such as the frequencies of unilateral vs. bilateral cases in the population. He also estimated the likely number of tumors in the retinoblastoma population. The data fit a mathematical model provided he made a key assumption: for the disease to occur, both copies of a gene that is essential for proper development of the retina must be damaged.

His model of the disease was supported by data showing that inherited cases occur at younger ages than sporadic cases in which no family history of the cancer is evident. In individuals born with a *germline mutation*—those with a mutated gene derived from the embryo's DNA—a second mutation that damages the second copy of the retinoblastoma gene after birth (a *somatic mutation*) would lead to the development of the disease.

In sporadic cases, in which the individual is not born with a mutated gene, the mutation and inactivation of both copies of the gene would be necessary to cause the disease. The probability of two debilitating "hits," with one on each copy of the retinoblastoma gene, is far, far lower than the likelihood of a single mutational event that can lead to retinoblastoma in people born with a damaged gene already on board.[54] You would therefore expect the sporadic cases to occur at older ages than the inherited cases.

Knudson's model provided an astounding fit to the data. It readily explained, for example, why retinoblastoma studies (including their own at M.D. Anderson) were remarkably consistent with respect to the mean age of diagnosis for the two forms of retinoblastoma. The data showed that diagnosis of the cancer occurred at approximately 15 months of age for cases of the inherited form of the disease and about twice that, around 30 months, for spontaneous cases of retinoblastoma.

The calculations also showed that the inherited cases were likely to generate cancers characterized by the presence of multiple tumors in one or both eyes. The model also predicted that the likelihood of the presence of multiple tumors in the case of spontaneous retinoblastoma was exceedingly small. In both cases the clinical data were consistent with these predictions.

Knudson pointed out that his key assumption, which he called the *two-mutation hypothesis*, is consistent with existing knowledge on the mutational origin of cancer.[55] It was evident at the time that the initiation and progression of most diagnosable human cancers resulted from multiple genetic insults. Estimates ranged from three to seven hits on the genome as a requirement for most human cancers.

The data provided in Knudson's mathematical *tour de force* provided direct support for the hypothesis that some human cancers may result from only two hits on the genome. "The data presented here and in the literature are consistent with the hypothesis that at least one cancer, retinoblastoma, can be caused by two mutations."[56]

Interest in hereditary cancers accelerated following the publication of Knudson's landmark two-mutation hypothesis, leading to further evidence for the purported nature of cancers such as retinoblastoma. The observed deletion of a section of chromosome 13 in some retinoblastoma cells suggested that a gene on that chromosome may be responsible for the genesis of the cancer.[57]

Experiments performed in the 1980s provided further evidence for the two-mutation hypothesis in retinoblastoma. In these experiments, the cell fusion technique pioneered by Kohler and Milstein for the development of hybridoma technology was used to fuse retinoblastoma tumor cells with normal cells. As predicted by the two-mutation hypothesis, the resulting hybrid cells lost the rapid-growth phenotype of cancer cells.[58]

This result suggested that a gene coding for a protein that normally serves as a brake on cancer growth is damaged in the tumor cells, allowing for rapid and uncontrolled cellular proliferation. In the hybrid, the change in phenotype of the cell must result from the restoration of a copy of the normal protein coded on the single intact copy of the gene passed on to the hybrid from the normal cell. The restoration of this protein's function must therefore be responsible for the prevention of runaway cell growth. This outcome provided further evidence for Knudson's hypothesis that both copies of the gene coding for the Rb protein were damaged in cases of retinoblastoma.

The isolation of a gene on chromosome 13 that correlated with retinoblastoma demonstrated the role of the gene product of the retinoblastoma gene. This protein, called the *retinoblastoma (Rb) protein*, is directly involved in the regulation of cellular proliferation. The Rb protein was later shown to play a key role in controlling cell growth as a gatekeeper in pathways leading to progression out of G1 into the initiation of DNA replication during S phase.

Further work demonstrated the presence of other genes in various hereditary cancers with characteristics like that of the Rb gene. Specifically, active protein products from both copies of the gene in question needed to be absent for the cancer to manifest. In cases where a mutation was already present at birth, the second, disease-causing mutation is required to activate the disease process.[59]

The products of genes like Rb are responsible for caretaker and gatekeeper processes such as the maintenance of DNA integrity and control of the cell cycle. Such genes were called *anti-oncogenes* to indicate they have a protective action against *oncogenesis* (aka *tumor initiation, carcinogenesis,* or *tumorigenesis*). They were later renamed *tumor suppressor genes* (and their products, *tumor suppressor proteins*) in recognition that "they serve to regulate normal cell growth; their loss removes a critical constraint on proliferation that in turn can contribute to tumorigenicity."[60]

In any specific case, the loss of tumor suppressor function may or may not lead to tumor formation, such that the impact of tumor suppressor loss is cell-type specific. For example, the loss of functional Rb protein leads to retinoblastoma and a bone cancer called osteosarcoma but not to other forms of cancer, even though the Rb protein plays a role in growth control in all cell types. This is due to other constraints on cell growth present in some cell types that must also be breached for tumorigenesis to ensue.

It is important to note that the presence of a hereditary lesion in one copy of a tumor suppressor gene such as Rb does not guarantee the future occurrence of retinoblastoma. For this reason, the term "hereditary cancer" is a misnomer, as the presence of the damaged gene provides, more precisely, an increased likelihood of developing retinoblastoma later in life. Even with this genetic defect, further genetic damage is required to establish a growing tumor. In the absence of this damage, the individual bearing the mutated gene is a carrier of the defective gene, capable of transmitting it to the next generation.

On January 22, 1971, at the yearly State of the Union Address, President Nixon introduced the National Cancer Act. which somebody later dubbed "The War on Cancer." The name stuck, despite the fact this colorful and belligerent phrase did not appear in the bill itself.[61]

During the speech, Nixon declared, "I will also ask for an appropriation of an extra $100 million to launch an intensive campaign to find a cure for cancer, and I will ask later for whatever additional funds can effectively be used. The time has come in America when the same kind of concentrated effort that split the atom and took man to the moon should be turned toward conquering this dread disease. Let us make a total national commitment to achieve this goal."[62]

This was an appealing vision. At the time, interest in the cancer research community turned to the potential of viruses to cause cancer. The discovery of multiple tumor viruses that can induce cancer in animals suggested that viruses might do the same in people.[63]

Ironically, the decades of groundbreaking discoveries about human cancer that followed Mr. Nixon's announcement resulted from, in retrospect, a rudimentary understanding of cancer's phenomenal complexity. Based on the evidence at the time, and because of how frightening it was to think that viral infections might flip a biological switch turning our own biology against us, the focus on viruses is certainly understandable when taken in historical context.[64] Fortunately, the scientific harvest from this research was exceedingly rich, if not somewhat misdirected from a conceptual standpoint.

There was, on the other hand, an upside to the dreadful prospect that viruses cause cancer. If viral infections caused most cancers, might it be possible to vaccinate the population against these viruses to stem the tide of this devastating illness?

As it turns out, the answer for most cancers is no. However, tumor incidence was significantly reduced for cancers induced by infection with two human viruses: *hepatitis B virus* (*HBV*)[65], which can cause liver cancer, and *human papillomavirus* (*HPV*), which can cause genital, anal, and oral cancers. While infection with these viruses does not directly cause cancer by mutating the host genome, the chronic inflammation they induce can lead to tumors over time. Vaccination is highly effective at reducing the incidence of human infection by these viruses and the cancers associated with them.

In the early 1970s, investigators focused on a monkey tumor virus called *Simian Virus 40* (*SV40*) as a model system for understanding virally induced tumorigenesis. SV40 can transform monkey cells in culture from the normal to the cancerous state. The virus can also cause cancer in monkeys. In addition, SV40 has a small and well-characterized genome. It is capable of infecting human cells, as well.[66] Finally, there has long been (difficult to confirm) evidence suggesting that SV40 may be associated with some human cancers.[67] These characteristics made SV40 an ideal virus for study.

During SV40 infection, two proteins, the *large T antigen* and *small T antigen*, are expressed by the virus during transformation of the host cell into a cancer cell. Biochemical analysis revealed the presence of a protein with a molecular weight of about 53,000 Daltons that was associated with the large T antigen in SV40-infected cells. The same protein was also found in uninfected mouse cancer cells and several other types of cancer cells.

These studies showed that the 53,000 Dalton protein, dubbed *p53*, was derived not from the virus but rather from the host genome. For some reason, this host protein was interacting with the SV40 protein during viral infection, suggesting that this host protein was somehow involved in transformation of the cell. Subsequent experiments showed that p53 participates in the transformation process. While it was appreciated at the time of the p53 discovery that the presence of a host cell protein in association with a viral protein offered an interesting clue to cellular transformation, the true significance of p53 in human cancer was not fully evident at the time of its discovery.

It took over two decades of dedicated research to unveil the scope of the role of the p53 protein in maintaining homeostasis. A 2017 publication surveyed the National Library of Medicine's publication database, called *PubMed*, and found that p53 was "the most studied protein and gene in literature, with a total of more than 80,000 entries."[68]

As a major player in the surveillance of DNA quality, p53 can detect DNA damage and activate biochemical signaling networks that impact the state of the cell. Through binding to both DNA and a multitude of protein partners, p53 determines whether a cell with damaged DNA should proceed through the cell cycle to DNA replication and mitosis or, alternatively, whether the cell should undergo an alternative fate. The cell might arrest in the cell cycle in the G0 state (*quiescence*), a potentially reversible (pending DNA repair) resting state; it might enter *senescence*, an irreversible step toward the end of the cellular lifespan when the cell can no longer divide (this is also known as a *post-mitotic state*); or the cell may undergo apoptosis, programmed cell death.

The p53 protein is the most critical monitor of cellular stress in the cell. It ensures that cells with damaged genomes do not enter DNA replication but rather are stalled in the cell cycle while the biochemical machinery assesses the appropriate future path for the cell. In the transformation of host cells by cancer viruses such as SV40, the viral transformation involves the diminution of p53 activity via the binding of p53 to the large T antigen. The inactivation of p53 results in deficits in DNA repair and cell cycle control that alter the genetic fate of the infected cells. As a result, productive members of the host's cellular community are forced through a series of transformations that turn them into cancer cells.

We now know that p53 mutations are found in about half of all human cancers. In some cancers, like colon cancer and head and neck cancer, mutated p53 appears to be involved about 70-85% of the time.

Because of its indispensable role, p53 has been dubbed the "guardian of the genome" and the "master tumor suppressor" of the cell. If we direct ourselves back to the DNA airport, we will find p53 in the chair of the shift supervisor at air traffic control. The critical importance of the shift supervisor is clear. If the deployment of resources and decisions at DNA air traffic control is no longer successfully managed and implemented, chaos can result.

As a rule, the loss of the functionally active gene product from both alleles of a tumor suppressor gene is required for the complete loss of tumor suppressor function. The p53 protein is an exception, as it is possible, due to p53's complex biochemistry, for a mutation in one copy of the p53 gene to result in the total loss of tumor suppressor function.

Amongst its many functions, p53 is a DNA-binding protein that activates multiple gene expression programs. Such a protein is known as a *transcription factor (TF)*. When p53 interacts with the DNA, it is not in the form of a single p53 molecule, but rather, it exists as a tetramer—four copies of the p53 protein operating together to elicit biological activities at the level of the DNA to control the expression of the genes.

Since p53 operates as a tetramer, its overall structure is determined by the interactions of the four individual p53 proteins that comprise it. A mutation that results in a structural change to p53 can impact the binding of the mutated form to copies of undamaged (*wild type*) p53. As a result, the overall structure of the tetramer can be sufficiently perturbed to reduce, or even eliminate, the binding of the tetramer to the DNA. The result of this loss of DNA-binding activity by the p53 tetramer can lead to the loss of tumor suppressor function and the acceleration of cancer progression, all resulting from as little as a single mutation that impacts the structure and function of one of the protein copies in the p53 tetramer.

The mechanism of action of p53 precludes the correction of p53 malfunction by the injection of normal p53 protein into the patient, as it does not eliminate the presence of the mutant p53 that is impacting the biological activity of the tumor suppressor in its functional tetrameric state.[69]

If only biology were so simple.

Chapter 4

Deciphering the Source Code

Our DNA exists in a highly compacted state, where the DNA helices are associated with complexes of proteins called *histones*. The DNA-protein complexes form fibers called *chromatin*. The chromatin fibers—which, taken together, form our chromosomes—are so highly compacted in the nucleus of our cells that approximately three billion DNA base pairs are present in a single human cell. If you took the DNA out of a single cell and uncoiled it, it would stretch about two meters, approximately six and a half feet.[70] Thus, the DNA from a single human cell stretched end to end is about as tall as Michael Jordan. Incredible.

But wait, there's more. If you took all the DNA from a single human's 30 trillion (or so) cells and uncoiled it, then set it end upon end, it would stretch about 60 trillion meters, which is 60 billion kilometers (about 37 billion miles). A round trip from Earth to the sun and back is about 300 million kilometers (about 186 million miles). Thus, the total DNA in our bodies, stretched end to end, would go to the sun and back about 200 times.[71] And if that's not sufficiently impressive for you, how about this: the DNA inside of us would extend from our planet to Pluto and back about four-and-a-half times.[72]

Here's a gross understatement: That's a lot of DNA. So how is all that DNA packed inside the cell?

The appearance of our DNA, when viewed under the electron microscope, has been described as analogous to "beads on a string." This appearance is due to the presence, at regular intervals along the DNA strand, of a protein "bead" around which the DNA is tightly wound. The bead is comprised of a double-helical strand of 146 base pairs of DNA wound around a protein complex of eight histone molecules, which collectively form the *histone core particle*. The combined DNA-protein structure, called a *nucleosome*, is shown in the upper right of Figure 7.

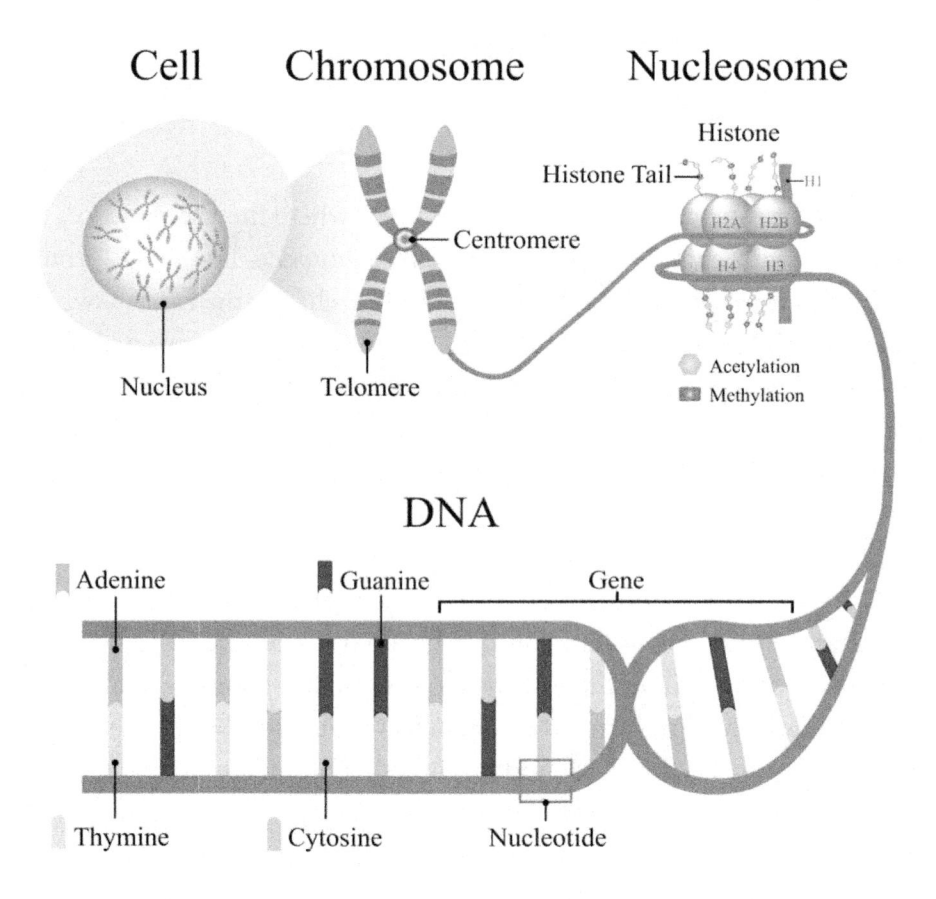

Figure 7. The structure of DNA in the human cell

The eight histones that comprise the core particle contain two copies each of four distinct, but closely related histone proteins known as H2A, H2B, H3, and H4. The region between core particles contains 54 DNA base pairs. Between the nucleosomes, histone H1, known as the linker histone, modulates the structural dynamics of the nucleosomes by impacting the flexibility of the linker.

Under the electron microscope, the nucleosomes form a well-ordered structure. The nucleosome strings are highly compacted together, forming the chromatin fibers that comprise the familiar x-shaped chromosomal structure visible under the microscope. However, in the living cell, the picture is far different: the two-dimensional photograph does not display the structure's three-dimensional complexity and dynamism.

In living cells, the chromatin fiber is in a constantly changing environment. The expression of the genetic program requires an intricate molecular dance that involves rapid and dynamic changes to the chromatin structure. These changes, which modify the chromatin's three-dimensional shape and level of compaction, alter the interactions of the proteins, RNA, and DNA that participate in the expression of the genes. The structural dynamics of chromatin are mediated by the action of dozens of enzymes that transfer small chemical groups onto (and off) both the histones and the DNA, driving changes that regulate the access to the DNA surface of the biochemical molecules involved in gene expression.

The discovery of the nucleosome's structure in the 1960s, which revealed how DNA is packaged in the nucleus, prompted further research to understand the biochemical mechanisms that control chromatin's structure and function. Recognizing that both the histone and DNA components of the nucleosome are subject to enzymatic modifications impacting their structures, the effort focused on determining the functions of these enzymes. By unraveling their roles in gene expression, the purpose of the enzymatic modifications of the histones and DNA was revealed.

By the mid-1970s, it was clear that chemical modifications correlate with the activity (or lack of activity) of the genes in the regions where the modifications are found. The data showed that a specific enzymatic modification of the histones is present in regions of active genes. This modification (called *histone acetylation*) involves the addition of a small chemical group to the amino acid lysine in a histone protein. In the acetylation reaction, the addition of just a few atoms to a histone protein molecule containing thousands of atoms results in a structural change that disrupts the binding of the histone protein to the DNA.[73] The loosening of the interaction between the histone and the DNA provides access along the DNA strand for the macromolecular machinery responsible for the expression of the genes.

An additional modification found on the amino acid lysine of the histone molecules, called *histone methylation*, involves the addition of a methyl group (CH_3) to a lysine on a histone. While histone methylation has functional implications, these effects vary depending on circumstances, sometimes activating gene expression, at other times, dampening it.

These histone modifications, orchestrated by enzymes with names like histone acetyltransferase (for the acetylation reaction) and histone methyltransferase (for the methylation reaction) take place at specific lysines on each type of histone protein in a manner that is both highly dynamic and reversible due to the action of other enzymes that remove the modifications.

The reversibility of these enzymatically driven modifications is a widely used molecular motif in nature. Evolution has ensured that the status of many biochemical reactions can be rapidly reversed when necessary. As the histones undergo these chemical modification reactions, the DNA's topography is altered, thereby providing a means of controlling genetic activity by dynamically regulating the shape of the chromatin in three-dimensional space.

In addition to enzymatic modifications of the histones, another chemical modification occurs on the DNA itself: the addition of a methyl group (CH_3) to the DNA sequence at specific locations, a modification known as *DNA methylation*. As a result of DNA methylation, the genes in proximity to the methylated region are rendered inactive, as the presence of the methyl group on the DNA prevents the binding of components of the gene expression machinery to the DNA. The reverse reaction, *demethylation*, involves enzymatic removal of the methyl groups from the DNA, which activates gene expression in that region of the genome.

The biochemical modifications of the histones and DNA, with concomitant effects on gene expression, are called *epigenetic modifications*. This term refers to changes in gene expression that occur without a change in the genetic sequence itself. The epigenetic modifications of the histones and DNA act together to dynamically shape and mold the chromatin to provide exquisitely precise control of gene expression.

During gene expression, the protein and DNA modifications result in changes in the three-dimensional structure of the impacted region of the chromatin which, in turn, impacts the accessibility of the genome to the metabolic machinery responsible for gene expression. In addition, these biochemical mechanisms lead to rapid rearrangement of the nucleosome locations along the DNA, exposing or blocking regions of the genome and thereby further regulating gene expression.

The enzymatic reactions responsible for epigenetic modifications occur in a rapid and coordinated fashion, courtesy of the precision and specificity of the molecular signaling mechanisms and enzymatic reactions present in biological systems. The structural alterations at the level of the chromatin are similar conceptually to the dynamics of protein structure, as proteins also alter their three-dimensional structures to modulate their biological functions.

Epigenetic modifications are highly orchestrated and carefully regulated to ensure the proper functioning of the genome throughout the lifespan. The patterns of chemical modification on the chromatin in fetal tissues, for example, are markedly different from those in adult cells. Similarly, cancer cells have distinct differences in epigenetic patterns compared to their normal counterparts of the same cell type. This observation suggests that these epigenetic differences play a role in cancer progression, by altering the patterns of gene expression.

Due to the evidence supporting the role of epigenetic changes in tumorigenesis and subsequent invasion and metastasis, research in epigenetics is a booming field, with thousands of peer-reviewed journal articles appearing monthly.

Following the elucidation of the DNA structure, the next step was to figure out how the information contained in DNA sequences orchestrates the activities of our cells. At the 1957 meeting of the Society of Experimental Biology on the nature of gene expression,[74] Francis Crick proposed a model for the relationship between genes and proteins that he called the *Central Dogma of Molecular Biology*.[75]

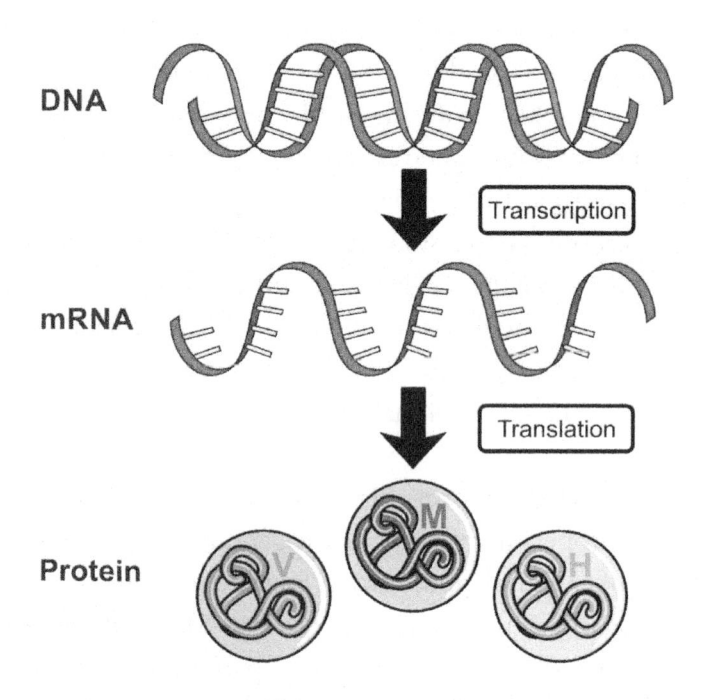

Figure 8. The Central Dogma of Molecular Biology

The fundamental concept encompassed in the central dogma is that the genetic source code stored in the DNA specifies the order of the protein's amino acid building blocks.[76] In the scheme envisioned by Crick (shown in Figure 8), the information in DNA encodes information for the production of an RNA in the nucleus of the cell. The RNA, in turn, transfers the genetic code from the nucleus to the site of protein synthesis in the cytoplasm.

Both DNA and RNA are comprised of strings of *nucleotides* twisted in a helical structure (like a corkscrew). In DNA, two helical strings are intertwined in a double helix, while RNA is in the form of a single helix. The DNA molecule contains two structural components: the "rails" of the ladder-like structure, known as the sugar-phosphate backbone, and the rungs of the ladder, the nucleotide bases that carry the genetic message (Figure 8). A nucleotide is comprised of the nucleotide base linked to the sugar-phosphate backbone—a *phosphate group*, which is a phosphorus atom surrounded by four oxygens—linked to the five-carbon sugar deoxyribose (in DNA) or its chemical cousin, ribose (in RNA).

The nucleotide bases found in DNA (as noted in Chapter 3) are called adenine, thymine, cytosine, and guanine, known by the letters A, T, C, and G, respectively. In RNA, thymine is replaced by the structurally related nucleotide base uracil (U). Watson and Crick's DNA model demonstrated that A always pairs with T (or with U in RNA), and G always pairs with C. This base-pairing motif is the key to how the cell replicates and interprets the genetic code to create the proteins needed for its survival.

When Watson and Crick proposed the DNA structure, they realized that the presence of the complementary strands of the double helix defined by the base-pairing rules suggested a means for DNA self-replication. Base complementarity allowed each strand to serve as a template for the creation of a partner strand. By complementary, we mean that if there is an A on one strand of the DNA, there is a T at that position on the partner strand; a C on one strand of the DNA dictates that there will be a G on the partner strand.

Crick proposed that just as each DNA strand can serve as a template for the synthesis of the partner strand in replication, a sequence on a strand of DNA is also used as a template during gene expression. During the process of *transcription*, a single-stranded RNA is assembled from a string of nucleotides using one of the DNA strands as a template that directs the creation of a complementary sequence, where A on the DNA pairs with U on the RNA (and T on the DNA pairs with A on the RNA), while G pairs with C.

The RNA product of transcription leaves the nucleus and goes into the cytoplasm of the cell. In the cytoplasm—all the material outside the nucleus and inside the cell membrane—the single-stranded RNA undergoes a process called *translation*. During translation, the RNA's nucleotide base sequence specifies a sequence of amino acids that are strung together to form the protein coded by the gene that specified the RNA. Translation converts the biological information from the language of the nucleotide bases into a string of amino acids that forms the protein specified by that gene.

At the time that Crick formulated the central dogma, the biochemical mechanisms behind the processes he described were completely unknown. Crick's proposal, while visionary and ingenious, was not, however, mere speculation. Rather, it was firmly based on the biochemical information available to him in the late 1950s.

Existing evidence showed that RNA resided in both the cell's nucleus and the cytoplasm. This suggested that RNA serves functions in both subcellular compartments. In addition, the data showed the presence of particles in the cytoplasm that contained both nucleic acids and proteins. This, presumably, was the place where RNA and proteins interacted to translate the genetic information carried by the RNA into a protein sequence.

These cytoplasmic nucleic acid-protein particles were discovered using centrifugation, a process that exposes cells to high-speed rotation in a rapidly spinning device called a rotor. Centrifuges were first used in the nineteenth century for separating the components of raw milk. You may be familiar with the concept of centrifugation from old films of astronaut training, in which prospective cosmic adventurers were subjected to centrifugal (outward-directed) forces of up to about seven times the force of gravity to determine who had the "right stuff" for the rigors of space flight.

The 1974 Nobel Prize in Physiology or Medicine recognized the pioneering work of Belgian cell biologist Albert Claude who used centrifugation to explore cell structure. He shared the prize with two other scientists for their work on the organization of the cell. Claude's work provided the first evidence for the presence of subcellular particles that contain a high concentration of RNA associated with proteins. These isolated particles, first called *microsomes* and later *nucleoprotein complexes*, contain balls of RNA and protein that function as factories for manufacturing proteins. Biochemists and cell biologists settled on the name *ribosome* for the RNA-containing microsomal particles where protein synthesis takes place.

The ribosome is a biochemical structure that operates as a protein synthesis machine. The ribosome is comprised of RNAs called *ribosomal RNAs* (abbreviated *rRNAs*) complexed with a complement of proteins that work with the rRNAs to accomplish the critical task of translation. During translation, the nucleic acid sequence on the RNA transcript is used as a template for assembling the string of amino acids that comprises the protein sequence encoded in the DNA.

Crick hypothesized in the central dogma the existence of a form of RNA that serves as a "messenger" for transferring the DNA code to the protein synthesis machinery in the cytoplasm. He proposed that the messenger was the product of the transcription of a gene on the DNA into a single-stranded RNA molecule.

Crick's hypothesis was confirmed with the discovery of *messenger RNA* (*mRNA*) in 1961, the result of experimental work performed by Crick, South African biologist Sydney Brenner, American biochemists Matthew Meselson and Arthur Pardee, and French biologists Jacques Monod and Francois Jacob. This form of RNA, the carrier of the genetic message, was (appropriately) called messenger RNA (mRNA*)*. Though imagined by Francis Crick in 1957, it was impossible to isolate the message until tools were developed to achieve the required experimental work. As a relatively short-lived macromolecule, mRNA is a challenge to isolate from living cells.

Ribosomes are the sites of protein synthesis in all living cells on Earth. No cellular organism—regardless of its taxonomical classification—

synthesizes proteins in any other way. The ribosome is so effective at making protein molecules that it is used as the mechanism for protein synthesis throughout the living world (Figure 9).

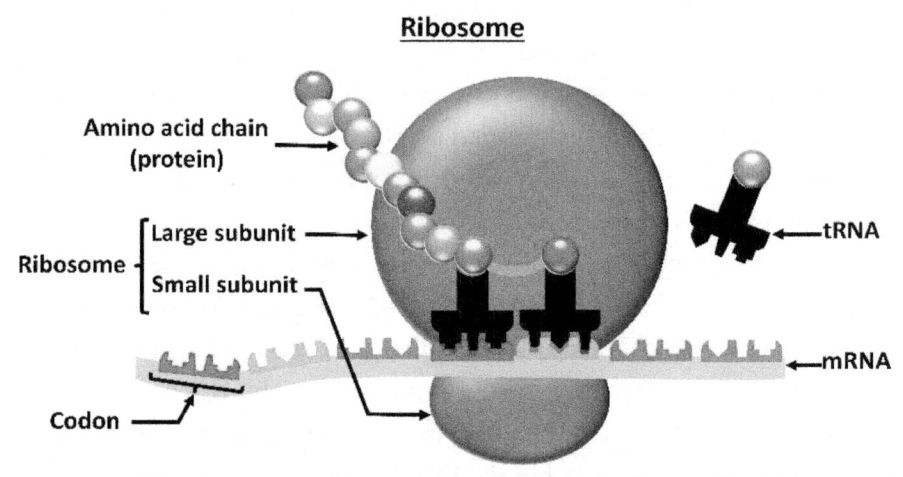

Figure 9. The structure of the ribosome

Once the mRNA arrives at the ribosome, the process of translation begins. Operating like a tape reader, the ribosomal structure engulfs the RNA messenger molecule between its two subunits and feeds it through the structure like a cassette tape threading through tape heads. Crick hypothesized that a molecule was required to bring the right amino acid to the ribosome as dictated by the genetic information on the message. Crick called this molecule an *adaptor* in recognition that its function was to link, or adapt, the code on the message to the specified amino acid.

Crick proposed there was an adaptor molecule for each type of amino acid, each with the job of bringing the correct amino acid to the ribosome in the appropriate order for assembling the protein. Hence, the adaptor molecule was an intermediary between the language of nucleic acids and that of amino acid sequences, a translator from one form of biological information encoded in the nucleotide bases to another form of biological language, that of amino acids, the building blocks of proteins.

Crick's hypothesis did not specify the adaptor's biochemical nature, though he recognized that the adaptor had the same type of code as that on the mRNA (and the DNA). The language of genetics, he believed, would provide a solution to how the correct adaptor might read the nucleic acid sequence on the message. This meant that RNA was a suspect for the job of the adaptor molecule.

The adaptor molecule—discovered by Mahlon Hoagland and Paul Zamecnik in 1958 at the Massachusetts General Hospital, shortly after the publication of the central dogma, was indeed comprised of RNA. It was called *transfer RNA (tRNA)* in recognition of its role in transferring the correct amino acid to the growing polypeptide chain at the ribosome.

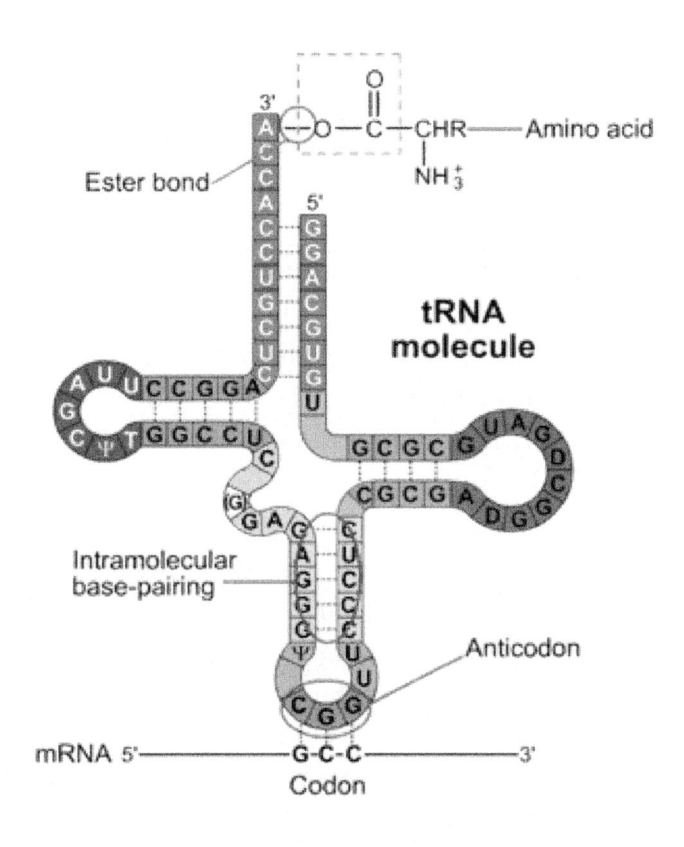

Figure 10. The structure of tRNA

The structure of the transfer RNA molecule (Figure 10) was worked out by Alexander Rich at MIT in the early 1970s using the technique that was also used for determining the structure of DNA and proteins: x-ray crystallography.[77] The transfer RNA sequence contains stretches of both unpaired RNA (called *loops*) and stretches of base-paired RNA, where complementary bases are hydrogen-bonded using the base-pairing rules to provide stability for the overall three-dimensional shape of the tRNA molecule.

At one end of the structure (at the top of the figure) is a short nucleotide sequence that is recognized by an enzyme carrying the amino acid specified for that tRNA. There is a specific enzyme for each amino acid and its tRNA. The enzyme transfers its bound amino acid to the nucleotide at the top of tRNA chain (Figure 10), forming a bond (called an *ester bond*) between the amino acid and the tRNA. This reaction, catalyzed by enzymes called *aminoacyl-tRNA synthetases*, is known as *charging* the tRNA with its amino acid.

At the other end of the tRNA structure (bottom of the figure) is a three-nucleotide sequence that binds a complementary three-nucleotide sequence on the mRNA as the message feeds through the ribosomal "tape reader." The three-nucleotide sequence on the tRNA, called the *anti-codon*, can bind to the *codon*, a complementary three-nucleotide sequence on the mRNA that is, in turn, a complementary sequence to that present on its DNA template.

For example, a DNA sequence of AGC is transcribed to an mRNA sequence of UCG. That codon sequence on the mRNA will bind, in turn, to a transfer RNA with the anti-codon sequence AGC (the same as on the DNA). The UCG codon specifies the amino acid serine.

This three-nucleotide code is used to transmit the information from the DNA to the ribosome, orchestrating the addition of the correct amino acid at each step of protein synthesis as encoded on the DNA. This three-nucleotide codon/anti-codon motif is the basis of the genetic code, the translation of the language of nucleotides to that of amino acids.

The experiments of Marshall Nirenberg, a biochemist at the National

Institutes of Health, provided the key to deciphering the code. In 1961, Nirenberg showed that when an RNA containing only the nucleotide uracil (U) was added to an experimental protein synthesis system, a polymer (repeated units) of the amino acid phenylalanine was produced. This meant that repeats of the triplet code UUU as the codon on the mRNA will produce a string of phenylalanines. Thus, UUU is the RNA code for phenylalanine, and its complementary sequence AAA is the DNA code for phenylalanine. Subsequent experiments revealed the remainder of the codons representing all 20 of the naturally occurring amino acids.

Francis Crick predicted that combinations of three nucleotides, which provides 4^3, or 64, combinations amongst the four possible nucleotides at each position, would be sufficient to code for the 20 naturally occurring amino acids. A two-letter code would only have 4^2 (16) combinations; a four-letter code would provide 4^4 (256) combinations, far more than needed. Having faith that the code would be efficient, Crick settled (as it turns out, correctly) on the prediction of a three-digit coding format.

The more commonly used amino acids in proteins have more codons than the less commonly used amino acids, and these codons share closely related sequences for a given amino acid. For example, the amino acids glycine, leucine, and serine, all commonly found in abundance in proteins, can each be specified by four different codons, whereas the less abundant amino acids phenylalanine and tyrosine are only encoded by two triplet codons each. This redundancy reduces the likelihood that single base changes will alter the amino acid sequence.[78]

The code's efficiency is also evident in the sparse use of additional codons beyond those needed to code for the 20 amino acids. There is a single "start" triplet that signals the beginning of a protein sequence and three different "stop" codons signaling that protein synthesis is complete.

It is therefore possible to look at a DNA sequence and determine where proteins start and stop. However, it is impossible to examine the nucleotide sequence and determine the amino acid sequence of the protein synthesized from a given "start" codon. This is because proteins are not encoded on the DNA in a simple linear sequence of nucleotides. Rather, the protein-coding sequences are interspersed with additional nucleotide sequences that do not code for amino acids.

After the mRNA transcript is synthesized, the non-protein-coding DNA sequences are edited out of the mRNA transcript in a process called *splicing*. The sequences that are edited out are called *introns*; the codons present in the final mRNA transcript that encode the amino sequence of the protein are called *exons*. It is important to note that a single unedited transcript can be spliced in multiple ways, such that multiple proteins can be made from a single RNA transcript.

Since the genetic code is based on a three-nucleotide motif coding for each amino acid, there are three ways to read a DNA sequence. These are called *reading frames*. For example, in the sequence AACTGGTAG, we can begin reading codons from the first A, or the second A, or the C in the third position. In any given DNA sequence where a protein can be made—that is, one with a start codon—there may be other start codons in other reading frames in that sequence. Accordingly, a single sequence can be transcribed into different transcripts depending on the nucleotide sequence in each reading frame.

Over the years, we have come to appreciate both the simplicity of the genetic code and the startling complexity embedded within it. It is a testimony to Crick's sweeping intellect and genius-level insight that his central dogma proposal has required no substantive modification in the six-and-a-half decades since it was first presented.[79]

Ironically, Crick's genius in this case did not apply in naming his theory, as his use of the word "dogma" is inappropriate in this context. A dogma is an authoritative statement considered beyond questioning. That was certainly not Crick's intent. When asked years later why he had called his proposal the "Central Dogma of Molecular Biology," Crick admitted that he didn't know at the time what a dogma was and chose that title because it sounded good.

By the 1960's, it was clear that while the DNA codes for protein sequences, at least some of the DNA sequences must code for the cellular complement of ribosomal and transfer RNAs. In addition, the data showed that binding sites for proteins regulating gene expression are scattered throughout the genome. Finally, the number of sequences on the DNA that serve as binding sites for transcription factors and other molecules involved in gene regulation, which were called *regulatory genes* in the bygone pre-genomic era, could not even be estimated. There was no way to determine the extent of the regulatory network encoded in the genome.

In the late 1970s and early 1980s, a popular debate amongst biological scientists interested in genetic matters focused on the question of why there was so much DNA in a human cell. The data suggested that even if there were hundreds of thousands of proteins in the cell, the genome had far more coding capacity than required. One school of thought was presented in a 1972 paper entitled "So Much 'Junk' DNA in our Genome." The paper's author, evolutionary biologist Susumu Ohno, argued that the percentage of the genome containing the "protein code" must have an upper limit.[80]

Ohno based this hypothesis on an evolutionary argument. If, he reasoned, the coding sequences became too abundant, the likelihood of a deleterious or fatal mutation would become, at some point, too great a burden on the cell to maintain a stable genome. He started with an assumption that a human cell had about three billion base pairs.

This estimate was based on a calculation involving several variables, starting with the experimentally determined weight of DNA in a human cell—about seven picograms, or seven trillionths of a gram.[81] From there, one can estimate the size of the double helix by using the molecular weights of the nucleotides and the sugar-phosphate backbone. Amazingly, Ohno's estimate of the genome's size was remarkably close to the estimated 3.1 billion base pairs provided by modern genomic data.

Once Ohno arrived at his estimate for the genome's size, he evaluated published data on the frequency of mutation of several protein-coding genes. He estimated that the mutation rate of the human genome is approximately 10^{-5} mutations per gene per generation. In other words, Ohno proposed that the likelihood of a mutation at each gene following DNA replication was about one in a hundred thousand.

In 1972, Ohno could not estimate the number of protein-coding genes. If there were 100,000 protein-coding genes, there would be about one mutation in the genome following each round of replication.[82] Noting that this mutation rate "appears to represent an unusually heavy genetic load"—meaning that, in such a case, the species might not survive—he concluded that the number of genes must be below this number.[83]

He concluded his argument by proposing that the genes are separated by long stretches of DNA sequences that are not transcribed into RNA, and "appear to function in a negative way," ascribing to these sequences "the importance of doing nothing."[84]

The reasoning behind this idea ran as follows: if a mutation is a random event, an excess of inconsequential sequences relative to the number of protein-coding sequences would reduce the likelihood that the "important DNA" would suffer from random mutations. From an evolutionary standpoint, Ohno was drawn to the concept that vast DNA sequences served as decoy targets for DNA-damaging agents to protect the protein-coding genes from harm.

He described these non-coding sequences as "partitions" that can dampen the potential impact of mutations in the active genes. He noted, using an argument based on natural selection, that these sequences may have once contained coding genes that are now dysfunctional: "The earth is strewn with fossil remains of extinct species; is it a wonder that our genome too is filled with the remains of extinct genes?"[85]

The potential presence of regions of the genome that are "doing nothing" with respect to the transmission of genetic information was dubbed the *Junk Hypothesis.* The idea embedded in the Junk Hypothesis—that the genome is composed of protein genes interspersed with large segments of non-coding, inconsequential DNA—was actively debated when I entered graduate school in 1977. From the time the Junk Hypothesis first appeared in the literature, there was a healthy current of skepticism in the scientific community about it.

While the basic idea of protecting the coding sequences from mutation certainly held great appeal, the scenario described by the hypothesis seemed horribly inefficient. The significant amounts of biological energy required to replicate and maintain so much "junk" raised obvious questions about why so much effort would be made to keep these partitions in place. Not only would this requirement drain cellular energy supplies, but it would also consume substantial amounts of macromolecular building blocks (such as amino acids and nucleotides). To some, that seemed a high price to pay for functionless "decoys."[86]

Opponents of the Junk Hypothesis proposed that biological systems could not afford to operate so inefficiently. They argued that the investment of energy and biochemical resources required to perform the vast numbers of biochemical reactions required to maintain genomic "junk" seemed too great for the purported purpose of protecting the genome. For the skeptics of the "junk" idea (I was in that camp), the actual purpose of all that DNA in the human cell remained a mystery.

The biochemical tools required to unravel the riddle posed by the Junk Hypothesis were not available until a quarter century after the publication of Ohno's proposal. By the dawn of the current century, a picture of the astounding complexity of the genome began to emerge that would forever change our view of the cell's inner workings and reveal the hidden secrets of the DNA code.

By the mid-1980s, three decades after the discovery of the double helix, biologists began to turn their attention to the grandiose ambition of sequencing the human genome.

The idea of pursuing the sequence of the entire human genome was first raised in 1985 at a meeting organized by molecular biologist Robert L. Sinsheimer of the University of California at Santa Cruz.[87] The potential significance of the effort was, as James Watson noted, "Similar to the 1961 decision made by President John F. Kennedy to send a man to the moon," predicting that "the implications of the Human Genome Project (HGP) for human life are likely to be far greater."[88]

Summing up the intensity of the interest in determining the sequence of the three billion or so base pairs of human DNA, Watson further observed, "A more important set of instruction books will never be found by human beings."[89]

The proposal was widely recognized in the scientific community as a compelling strategy for understanding the biochemical basis of life. As Watson put it, the sequencing of the human genome "will not only help us understand how we function as healthy human beings, but will also explain, at the chemical level, the role of genetic factors in a multitude of diseases—such as cancer, Alzheimer's disease, and schizophrenia—that diminish the lives of so many millions of people."[90]

When the project was proposed, the scientific tools required to pursue the program's objectives did not yet exist. It was, however, obvious that significant effort would be required to make the project's goals a reality. The DNA sequencing methods of the day, based on the pioneering sequencing work performed in the 1970s by Walter Gilbert of Harvard and Fred Sanger of Cambridge University, were capable of sequencing hundreds of base pairs per experiment.[91] This was far short of the throughput that would be required to sequence the human genome.

By the late seventies, complete genomic sequences were published for bacteriophages (viruses that infect bacteria) and the mammalian viruses SV40 and Epstein-Barr Virus. These genomes ranged from thousands of base pairs to hundreds of thousands of base pairs, many orders of magnitude smaller than the human genome.

It was evident at the time that further developments in sequencing methodologies, sequencing instrumentation and data handling were critical factors for the project's success. It was clear the project should be performed in stages, each stage building upon the technological achievements of those that preceded it.

The project plan called for the sequencing of genomes of increasing size and complexity to provide the experience necessary to tackle the ultimate prize: the human genome. Along the way, innovative technologies would be developed and improved over time to meet the sequencing throughput and data-handling requirements inherent in the effort to sequence the human genome. From the outset, it was reasonable to assume that the project might require at least 15 years, maybe even 20, for the completion of its lofty goals.[92]

The genomes chosen for sequencing, detailed below in order of increasing complexity, took advantage of the well-characterized model biological systems used globally for biological research: the bacterium *Escherichia coli*, the yeast *Saccharomyces cerevisiae*, the nematode worm *Caenorhabditis elegans*, the fruit fly *Drosophila melanogaster*, and the ever-useful laboratory house mouse *Mus musculus*.

President Ronald Reagan signed an appropriations bill for the National Institutes of Health human genome project in December of 1988, providing $17.2 million for the first fiscal year.[93] According to the second of two five-year plans issued by the NIH, a target was set by the Director of the NIH's Human Genome Research Institute, Dr. Francis Collins, to provide a draft sequence of the human genome by the end of 2001, and a complete and verified sequence by the end of 2003.[94]

The results proved the validity of the genome project's phased approach. It did not take long until the genomes of the non-human test organisms began to appear, one after another, fully available in the open literature for all to see. The draft sequence of the human genome was published in February 2001, and by April 2003, the complete human sequence was published, meeting the timeline established 15 years earlier at the project's inception. Along the way, there was an exponential growth in sequencing capability, made possible by the cooperation and collaboration of the many participating academic laboratories and research institutes, along with a healthy boost from private industry.[95]

The sequencing data from the genome project are freely available around the world, an open book that contains the source code that holds the key to human life. The technologies and scientific insights garnered by the effort have provided significant new tools for the investigation of human biology and disease; those, too, are fully described in the open literature.[96]

This is not to say the effort was devoid of controversy and interpersonal conflict. Scientists are human beings with strengths, foibles, personalities, moods, emotions, and egos. But in the end, the job was achieved by a global team effort, a little ahead of schedule and even under the assigned budget. If only governments could work like this.

In the same year that Robert Sinsheimer proposed the genome project, a sudden stroke of brilliance by a young chemist named Kary Mullis forever changed the biological sciences.

Mullis was a research scientist at a California biotechnology company called Cetus. Educated as a chemist at Georgia Tech and then in biochemistry at Berkeley, his area of expertise was DNA chemistry. His assignment at Cetus was to develop a way to reliably determine the presence of single base changes in long stretches of DNA.

These changes could be due to point mutations in the DNA, where one nucleotide is copied incorrectly during replication or damaged by environmental stress. These differences can also result from naturally occurring single base differences between individuals. Such naturally occurring base differences are known to biologists as *single nucleotide polymorphisms*, or *SNPs* (pronounced "snips").

Upon initiating this work, Mullis knew he was facing a challenging scientific problem. Using the sequencing technology of the age, it was not feasible to measure a single nucleotide difference in a reproducible way. There just wasn't enough DNA to obtain reliable measurements. You could try to amplify the signal in the analytical assay by using more sensitive detection methods; even then, you still had to amass a lot of sample to even make the attempt.

One warm spring day, while driving from Berkeley to the Mendocino coast with his girlfriend, he was thinking about his dilemma when he suddenly realized that he could use DNA as a template to make more copies of itself. As he thought about the idea, and his excitement grew, he began to wonder why no one had thought of it.

It seemed so obvious. Mullis thought this couldn't be an original idea. Or could it? As he noted in his 1993 Nobel Prize in Chemistry acceptance speech, "I thought, it had to be an illusion. Otherwise, it would change DNA chemistry forever."[97] When Mullis arrived at his getaway cabin in Mendocino, he put pen to paper for hours, staying up all night as he devised a chemical scheme that might provide a solution to the problem.

His idea was based on the well-established properties of nucleic acids, starting with the idea that the base-pairing between the complementary bases in the double helix can be interrupted by energy (in the form of heat). By heating a DNA sample, the hydrogen bonds that stabilize the complementary base pairs in the helix begin to come apart.[98]

At 95°C (just under the boiling point of water at 100°C), all the hydrogen bonds between complementary base pairs are broken, and the DNA strands come completely apart. The process of unraveling the structure of a macromolecule such as DNA (or a protein) is called *denaturation*, a term which indicates that the molecule is no longer in its natural biophysical configuration (conformation). In the case of DNA, the heat denaturation process that separates the DNA strands is known as *DNA melting*.

Separating the strands was the easy part. Given the complexity of DNA replication, the solution to the difficult part—copying the DNA sequences many, many times—was an entirely different matter. This is because the enzyme that copies the DNA, DNA polymerase, needs to attach each nucleotide to an existing nucleotide adjacent to it. The problem here was that DNA polymerase cannot start a new strand on its own. Rather, it needs a nucleotide to latch onto before it can extend the strand.

Evolution's clever solution to this dilemma is simple: RNA polymerases, the enzymes that create strands of RNA, can create brand new DNA strands. In DNA replication, an RNA polymerase called *RNA primase* creates a small anchoring piece of RNA (called a *primer*) that is complementary to the sequence on the DNA opposing strand. DNA polymerase then uses the primer as a starting sequence for the addition of nucleotides to the DNA strand. A different DNA polymerase enzyme subsequently replaces the RNA primers with DNA.

When Mullis returned to Cetus in Emeryville, he conducted a thorough literature and patent search. As he anxiously examined the literature, he could not believe that no one had beaten him to the punch. He realized that if he was right and the chemistry worked as he envisioned, he just might be on his way to something big.

He realized he would need to create primers to successfully replicate the DNA. Given his expertise in DNA chemistry, he believed it might be possible to use a small piece of single-stranded complementary DNA as a primer instead of using an RNA primer. The DNA primers made in the laboratory would base pair with known complementary sequences on each DNA strand, thereby serving as starting sequences for the DNA polymerase enzyme. By using DNA, and not RNA, there would be no need to replace the RNA with DNA, an additional complication that was best avoided in making copies of the DNA sample.

There were several potential pitfalls. First, he would need to find a temperature that allowed the primer to be hydrogen bonded to its complementary strand without reassociating the pre-existing complementary DNA strand with its partner strand. This was theoretically possible because the energy barrier for reforming all the hydrogen bonds in a long sequence is far greater than that for a far shorter sequence. It was, therefore, feasible that he might be able to find a temperature "sweet spot" where the primers would bind to the DNA without interference from the strand complementary to the one bound by the primer.

Next came the most challenging problem. The enzymatic addition of bases to the primer would obviously require an active DNA polymerase enzyme. Like many enzymes, DNA polymerases lose bioactivity at elevated temperatures. This is because the heat energy breaks some of the hydrogen bonds between the amino acids that hold together the protein's three-dimensional structure. When the biologically active structure is lost, a diminution (or total loss) of biological activity (function) follows.

You therefore could not use, for extension of the DNA strand, the DNA polymerase enzymes from sources such as *E. coli* or mammalian cells (mouse, human, etc.). These enzymes cannot function at temperatures where the two strands of the DNA double helix remain apart. Many enzymes in nature lose their structural integrity (and ability to function) above 50°C (122°F), or thereabouts. The temperature required to pull off the DNA copying scheme would be higher than most life forms can tolerate.

In evolutionary terms, there has been no selective pressure for enzymes that function at internal temperatures far higher than those of (most) living things.[99] In the face of this problem, Mullis's team arrived at an inspired idea: they realized that the diversity of nature shall provide. While a normal, everyday DNA polymerase was entirely unsuitable for the job, there were organisms that had developed under selective pressure to survive at higher temperatures.

The answer was found in a bacterial species called *thermus aquaticus*. *T. aquaticus* is an *extremophile*, an organism that thrives under conditions outside the typical ranges of the majority of life forms on Earth.[100] The DNA polymerase from *Thermus aquaticus* (called *Taq polymerase*) has evolved to operate at the temperature of the microbe's environment and is fully active at 55°C (131°F). Such an enzyme was perfectly happy at elevated temperatures.

Mullis believed that once the complementary DNA strand was fully synthesized by the Taq polymerase at 55°C using synthesized DNA primers as starting sequences, the temperature could be lowered back to the physiological range (37°C). Doing so should allow the complementary DNA strands to re-associate by hydrogen bonding between the base pairs. If this worked, the original piece of DNA could be used to create an identical copy of itself.

Theoretically, there was no limit to the number of rounds of DNA copying if the four DNA nucleotides were present in sufficient amounts to make multiple copies. There was reason to believe that millions to billions of rounds might be possible. In this manner, a small amount of DNA might be used to generate a pure DNA sample in sufficient quantities for dependable, high-precision sequencing.

Mullis and his team faced difficult technical hurdles. They were working with minuscule amounts of DNA, a challenging goal given the ever-present risk of contamination from the equipment and the scientists themselves. After months of demanding work, they were able to reproducibly generate pure DNA from human samples in the quantities required to reliably determine the single nucleotide changes at the core of their research objective.

The method of DNA amplification invented by Kary Mullis came to be known as the *polymerase chain reaction*, or *PCR* for short. Not only was Mullis able to solve the problem of determining tiny differences in human sequences between individuals, but the PCR technique soon became (and remains) an indispensable workhorse method in genetics and molecular biology laboratories worldwide. PCR is also the core technology of DNA forensics methods used in law enforcement and diagnostic testing for infectious diseases. These diagnostic tests include sensitive and accurate PCR tests for the presence of viruses such as the *human immunodeficiency virus (HIV)* and *severe acute respiratory syndrome coronavirus 2 (SARS-CoV-2)*.

PCR is an enabling technology for the biotechnology industry that provides the capability to create the genetic constructs needed for producing therapeutic proteins in cultured cells. This technique has played a key role in the discovery and manufacturing of a new generation of biomolecules made in living cells (biopharmaceuticals) that are used to treat grievous illnesses around the world.

The importance of Kary Mullis' discovery was recognized by the 1993 Nobel Prize in Chemistry, which he shared with British-born biochemist Michael Smith of the University of British Columbia-Vancouver.[101] Mullis' discovery provided the enabling technology for launching the era of genomic sequencing. It also provided a means for investigating the potential correlation of single nucleotide differences with many human diseases.

While cause-and-effect relationships between many SNPs and human diseases have not been established, and many of these single base changes have no evident impact on human health and disease, some single base changes have been directly correlated to specific diseases. However, as the saying goes, correlation does not reflect causation. These differences may reflect a variation in the predilection toward certain diseases; that is, they may enhance the statistical likelihood of contracting certain diseases amongst individuals in the population. If so, it is unclear what mechanism might be at play that provides for such outcomes. Whether these observed correlations have any direct relationship to the etiology of human diseases in cases where correlations have been observed remains an open question.

Kary Mullis' spectacular stroke of genius provided a biochemical invention that would soon become the scientific engine underlying the ability to sequence the vast nucleotide expanses of the human genome. For this reason, and those stated above, PCR stands as one of the towering scientific achievements of the twentieth century.[102]

Chapter 5

Living in an RNA World

The sequencing of the human genome provided a comprehensive view of the arrangement of the genes on our chromosomes, revealing a treasure map of how our genes are organized. The results confirmed the suspicion that most of the 3.1 billion DNA base pairs do not code for proteins. Rather, the genome's protein structural genes are separated by vast stretches of DNA that do not have protein-coding functions.

The results also showed that the protein-coding genes are commonly found in clusters, sometimes hundreds of genes long. The data showed there are approximately 21,000 protein-coding genes in the human genome, a surprisingly low level of about 1.6% of the total genomic sequence. This finding put the number of protein-coding genes at the bottom end of the range of previous "consensus" estimates of 15,000-150,000 genes.[103]

Given our protein-coding capacity is not much bigger than that of the nematode worm *C. elegans*—which has a tad under 20,000 genes in its genome—it was reasonable to assume that the vast stretches of non-protein-coding DNA in the human genome must do something interesting to functionally differentiate humans from roundworms.[104]

As the functional significance of the non-protein-coding portion of the genome—that is to say, most of the genome—was unknown at the time the protein-coding sequences were identified, it was as if the Human Genome

Project had found the treasure map, but less than 2% of it was legible. Though it was known that the genome also coded for the ribosomal and transfer RNAs, and that many sequences served as binding sites for DNA-binding proteins (such as transcription factors and DNA and RNA polymerases), many questions remained about the role of the genome's extensive non-protein-coding segments. Thus, a cavernous knowledge gap remained about how the genome functions.

This gap was addressed by a subsequent international research effort called *The Encyclopedia of DNA Elements (ENCODE)* project. ENCODE was designed to investigate the nature and, hopefully, the functions of the genome's non-coding segments. The project, described in a 2004 *Science* article from the ENCODE Consortium, sought to generate a "parts list" of the genome's functional elements beyond the relatively small number of protein-coding genes.[105] These functional elements include the observed RNA transcripts (known as *non-coding RNAs*) that do not appear to code for proteins, transcriptional regulatory sequences—that is, protein and RNA binding sites, and sequences that mediate chromatin architecture and dynamics.

A series of back-to-back publications in the September 6, 2012, edition of *Nature* magazine announced the findings of the ENCODE project. The lead article's abstract declared that the "project dishes up a hearty banquet of data that illuminate the roles of the functional elements of the human genome."[106]

The ENCODE findings drove a stake through the Junk Hypothesis. It provided a fascinating picture of the genome that revealed the true richness of the genetic code: the data showed that about 76% of the genome generated RNA transcripts for non-coding RNAs. Not only is the genome devoid of wide swaths of junk; most of the DNA sequences in the genome are transcribed into RNA. The DNA airport is far busier than we had ever imagined.[107]

In addition, the data indicated that about half of the genome can bind transcription factors and other macromolecules. It was even possible that most, perhaps even all, of the genome contained elements with biological functions related to the genome's regulation, including non-coding RNAs embedded in the introns of protein-coding genes. Although they are spliced out of the mRNA transcript before translation of the protein-coding sequences, the introns have biological significance.

The ENCODE papers provide a large-scale model of the genome in which the spaces between the protein-coding genes are laden with functional elements of distinct types. These elements include *promoters*, the sites of binding of transcription factors that initiate gene transcription, and *enhancers*, sequences that are functionally like promoters but far more powerful in their impact on gene transcription.

We can think of enhancers as "super promoters" that can initiate the transcription of large sets of genes, often at a significant distance from the enhancer site. This characteristic of "action at a distance" differentiates them from promoters, which operate locally on DNA sequences adjacent to the promoter's location.

The data also demonstrated that some of the known single nucleotide polymorphisms that appear to correlate with human disease occurrence are found within non-coding elements. This intriguing finding indicated that the investigation of genomic disease associations needs to extend beyond the protein-coding genes to all the non-coding RNAs. With respect to genomics and human disease, ENCODE brought forward the true scope of human genetic complexity.

For the scientists who cut their teeth in experimental biology around the time the Junk Hypothesis was proposed, the revelations brought to light by ENCODE about the genome's remarkable complexity were far beyond anything remotely imaginable by biochemists when I was training in the 1980s. With the ENCODE results, it was clear that protein production is regulated not only by transcription factors and DNA-binding proteins, but also by an immense network of RNA molecules. It is as if a computer dedicated over 98% of its memory to the operating system that orchestrates the tasks performed by the tiny percentage of the total memory capacity occupied by the software applications.

The ENCODE results posed new questions about gene expression:

Why is so much of the genome dedicated to its own control? How do these control mechanisms work? And finally, what is all this RNA for?

Biological scientists have long looked to model organisms to investigate the scientific principles of life. In the twentieth century, such models included the use of the fruit fly *Drosophila melanogaster* for the study of genetic mutation; bacteriophages as model systems for the study of gene regulation; and the common intestinal bacterium *Escherichia coli* as a model organism for the study of cellular physiology, metabolism, and the organization and regulation of the genes.

The common roundworm (nematode) *Caenorhabditis elegans* was first developed as a system for genetic study in the 1960s by Sydney Brenner. At the time, Brenner, a South African biochemist, shared an office at Cold Spring Harbor Laboratory (on the north shore of Long Island) with Francis Crick. In May of 1961, Brenner published a paper in *Nature* with two renowned molecular biologists, Francois Jacob and Matthew Meselson, entitled "An Unstable Intermediate Carrying Information from Genes to Ribosomes for Protein Synthesis."[108] The paper described the discovery of the messenger molecule predicted by Crick in 1957.

C. elegans is a free-living, non-parasitic nematode, a roundworm commonly found in soil. The adult animal, which grows to about a millimeter (about one twenty-fifth of an inch), comes in two genders: a self-fertilizing hermaphrodite and a male that is capable of mating with a hermaphrodite.[109] Brenner's interest in this organism came from the recognition that due to a peculiarity of its anatomy, the nematode provides an ideal model system for the study of genes and development: *C. Elegans* has a transparent outer cuticle through which it is possible to see the gonads, the reproductive organs, in the hermaphrodite.

A technique known as micro-injection that uses a tiny needle to place macromolecules directly into the worm's gonads was developed in Brenner's laboratory at the University of California at Berkeley. This technique provided the capability to place DNA, RNA, and/or proteins directly into the egg cells in the worm. When these biological molecules were placed directly into the eggs, they impacted the growth and development of the embryonic worm. Using this convenient experimental system, it was possible to pursue an understanding of the biological behavior of the injected macromolecules in a living organism.

Along with his colleagues H. Robert Horvitz and John E. Sulston, Brenner's research provided an understanding of how genes impact organ development and the biological process of programmed cell death, in which molecular signaling mechanisms mediate the destruction of defective cells by initiating apoptosis. As a result of this work, the three scientists shared a Nobel Prize in 2002. In his acceptance speech, which he called "Nature's Gift to Science," Brenner noted, "Without doubt the fourth winner of the Nobel Prize this year is *Caenorhabditis elegans*; it deserves all of the honour but, of course, it will not be able to share the monetary award."[110]

Unbeknownst to the Nobel Prize recipients that day, the common roundworm *Caenorhabditis elegans* had not made its final contribution to our understanding of genetic regulation. Sydney Brenner correctly predicted that the nematodes were not destined to accumulate wealth or fame. But their role in biological discovery would follow a similar path to that of the fruit fly, providing a platform for important biological discoveries that would reverberate for decades, if not centuries.

American biochemists Andrew Fire and Craig Mello were introduced to *C. elegans* as a model system for genetic studies in graduate school in the 1980s—the former at the Massachusetts Institute of Technology and the latter at the University of Colorado. After establishing their research groups—Fire at Stanford and Mello at the University of Massachusetts—both scientists continued using *C. elegans* and the techniques pioneered by Sydney Brenner and his group. This included the micro-injection of macromolecules in their studies of the regulation of gene expression.

Biologists believed at the time that gene expression levels could be elevated by increasing the amount of mRNA carrying the genetic information that specifies a given protein. Oddly, micro-injection of mRNA for a muscle protein (called *unc-22*) in the nematode did not increase the levels of the muscle protein at all. This result was odd, and it was a mystery how this could be. One possibility that came to mind was that an RNA sequence complementary to the mRNA for the unc-22 protein might be present. If so, it might base pair with the added mRNA for that gene, thereby rendering the single-stranded mRNA unsuitable as a template for protein synthesis.

In other experiments, the two RNA strands, the mRNA and its complementary RNA strand, were added together by micro-injection. Since the two different strands are complementary sequences, they base pair together when mixed, forming *double-stranded RNA* (*dsRNA*).[111] Amazingly, the presence of the double-stranded RNA led to a complete shutdown of the production of unc-22, resulting in a twitching phenomenon in the worm due to the lack of the muscle protein.

This was bizarre, not to mention extraordinary. Somehow, the presence of a dsRNA with a sequence on one strand complementary to that of an mRNA in the worm eradicated the production of the protein encoded by that mRNA. If the mRNA and its complementary sequence in the double-stranded RNA were bound together, as complementary sequences are inclined to do, how did the dsRNA eradicate the production of unc-22?

Mello called this phenomenon *RNA interference* (*RNAi*). Uncovering the mechanism of RNA interference required further investigations by Fire and Mello to determine the biochemistry behind this mysterious phenomenon. This investigation resulted in a 2006 Nobel Prize in Physiology or Medicine for the American biochemists for their discovery of "a fundamental mechanism for controlling the flow of genetic information."[112]

The path to Fire and Mello's discovery originated in the 1930s, when experiments on host responses to viral infections (such as yellow fever) provided some surprising results. If a virus is injected into a host animal (such as a rodent or rabbit) vulnerable to that virus, an infection is virtually guaranteed. Suppose the same animal, already infected with that virus, is subsequently infected by a second virus of a different type. In that case, the initial viral infection affords protection against infection by the second viral challenge. This is true even if the first viral challenge involves a viral strain that causes a mild infection, and the second challenge involves a far more virulent, or even lethal, virus. There is a marked protective effect to the second viral exposure.

This phenomenon was called *viral interference*. At the time, it was considered a kind of general protective response, as opposed to the case where an infected animal is re-challenged with the same virus. In the latter case, the immune system, primed by the first exposure to the virus, can mount a specific immune response against subsequent exposure.

The biochemical nature of viral interference remained unknown for over three decades. In 1957, a small protein called *interferon* was found circulating in the blood of patients with viral infections. Further work showed there are several types of interferons in mammals like us. In addition to a role in protection against viral infections, interferons are also important modulators of the immune response, members of a family of immune modulatory proteins called *cytokines*.[113]

While the genomes of microbes, animals, and plants are comprised of double-stranded DNA, viral genomes contain either DNA or RNA. DNA viruses contain double-stranded DNA, whereas RNA viruses can have genomes comprised of single- or double-stranded RNA. Apart from *retroviruses* such as HIV—which use an enzyme called *reverse transcriptase* to make a DNA copy of its single-stranded RNA, which is subsequently inserted into the host genome—many of the single-stranded RNA viruses reproduce their genomes via a double-stranded RNA intermediate.

Given the presence of double-stranded RNA in some viruses, the presence of RNA in this configuration inside of us indicates the possibility of a viral infection. The interference phenomenon induced by dsRNA is a response to the presence of the virus, an innate mechanism of viral defense. As Craig Mello noted in his Nobel Prize lecture, the discovery of double-stranded RNA as a mechanistic trigger of the viral interference response had uncovered "an ancient system by which cells could sense a molecule that was a bellwether of viral infection and respond by producing a signal that would tell the organism to dedicate its efforts and energies toward fighting viruses."[114]

While interferon is produced in response to viral infection, with profound effects on the immune system's anti-viral response, the biochemical ability to find and directly inactivate viral mRNAs belongs not to a protein but rather to a small piece of RNA dubbed *small interfering RNA (siRNA)*, the mystery molecule behind both viral interference and the phenomenon of RNA interference discovered by Fire and Mello in roundworms.[115]

When Fire and Mello discovered RNA interference in *C. elegans*, it was clear the nematode's unique properties provided an excellent experimental system for pursuing the biochemical mechanism of interference. By applying a tried-and-true, decades-old technique, it was possible to screen for worms deficient in RNA interference. By identifying mutants with deficiencies in conducting RNA interference, the genes responsible for RNAi were discovered, leading to elucidation of the components of the RNAi molecular machinery.

This work led to the discovery of the *RNA Inducible Silencing Complex*, known by the acronym *RISC*, which is the protein complex that processes double-stranded RNAs into small and highly targeted single stranded interfering RNAs. RISC is found in the cell's cytoplasm, where it awaits the arrival of a dsRNA precursor that it processes, courtesy of an enzyme called *argonaut*, into a mature single-stranded *small interfering RNA (siRNA)* that is 21-23 nucleotides long.

In association with RISC, the siRNA targets a complementary RNA sequence on a specific messenger RNA. Once RISC associates with a message that contains a sequence complementary to the sequence on the siRNA, the mRNA is either cleaved by the complex or rendered biochemically unstable and incapable of participating in protein synthesis.[116]

The interaction between the siRNA/RISC complex and the mRNA does not require the mRNA to bind a complementary sequence covering the entire length of the siRNA. Current data suggests that about two-thirds of the siRNA sequence may be required. This property allows an individual siRNA to interact with multiple mRNA messages, hundreds or more, using the base-pairing mechanism. Evidence from studying the genomes of vertebrates suggests that about half of all messenger RNAs in vertebrate species undergo regulation by this siRNA-mediated mechanism.[117]

The identification of other forms of functional RNA species that do not code for proteins followed the discovery of RNA interference. The discovery of an astounding variety of biologically active non-coding RNAs dramatically augmented the complexity of gene expression relative to what was envisioned by the pioneers of molecular biology.

Functional RNA species range from about two dozen nucleotides in length (these are called *microRNAs, or miRNAs*) to species that are several hundred nucleotides long (known as the *long non-coding RNAs*, or *lncRNAs*). These single RNA strands of approximately 200-300 nucleotides exert genetic control by interacting directly with specific transcription factors to modulate their binding to the DNA.

RNA-mediated control over the transcription factors impacts transcription factor behavior in profound ways. The presence of these RNA species augments the interaction of the TF with its binding site, turning it into a "super TF" that impacts large sets of genes, sometimes at a significant distance from its binding site on the DNA. Other members of this class of miRNAs can shut down a TF's activity by blocking the binding of the TF to its DNA binding site.

Recent work reveals another remarkable property of non-coding RNAs: Long non-coding RNAs can be transcribed from the DNA in both directions, with a different functional outcome depending on the direction of synthesis of the RNA. This finding shatters the long-held view that all transcription occurs in only one direction along the DNA strand, which is the case for the transcription of mRNAs, tRNAs, and rRNAs. It is likely that this mechanism is applicable for controlling many genes. This provides a highly efficient means for genetic regulation via the use of a single genetic element that can elicit opposing biological effects, thereby turning gene expression at a specific site on or off. Pretty nifty!

Another vital role for non-coding RNA is the cutting and splicing of RNA transcripts after they are synthesized. The use of RNA molecules for processing RNA transcripts is ubiquitous throughout nature, in organisms as disparate as humans, bacteria, fungi, and viruses. The Nobel Prize-winning work of Sidney Altman and Thomas Cech (1989, Chemistry) demonstrated RNA's startling versatility in the form of RNA molecules that can directly cleave precursor RNAs to generate mature RNA species, including the processing of messenger RNA transcripts into their processed bioactive forms.

The ability of RNAs to splice RNA indicates that, like proteins, RNAs can function as enzymes. RNA species with enzymatic activity have been called *ribozymes* in recognition of their ability to catalyze biochemical reactions such as splicing, with the specificity mediated by base pairing between the ribozyme and the RNA molecule it processes.[118]

An intriguing form of RNA that has been known for over a quarter century, but long dismissed as a byproduct of aberrant RNA splicing, has attracted attention for its recently demonstrated role in genetic regulation. Evidence indicates that *circular RNAs* (*circRNAs*) serve as "sponges" for the inactivation of miRNAs with gene-regulatory properties.[119] Circular RNAs can also bind to proteins that regulate gene expression. Abnormalities in the levels of these single-stranded circular RNAs have been detected in cancer cells, where they interact with miRNAs, such that circular RNAs are implicated in cancer initiation and progression.[120]

RNA's amazing versatility suggests it had a critical role in the evolution of life. It is believed by some investigators in molecular evolution that RNA was the first macromolecule, rather than strings of amino acids (proteins). The attraction of this hypothesis is two-fold. First, RNA has a vast array of functions, including reaction catalysis (enzyme-like activity that increases the rate of a chemical reaction), with specificity determined by simple base pairing. Second, RNA, unlike proteins, is capable of self-replication using the base-pairing mechanism, providing for the synthesis of a complementary strand based on the sequence of an existing strand. This is something proteins cannot do.

Synthesis of a complementary RNA would be extremely slow in the absence of enzymes like RNA polymerases. However, it is theoretically possible, given vast stretches of time, that an RNA strand could function as a template for the synthesis of a complementary strand by random collisions of nucleotides in the primordial soup. This line of reasoning has led to the concept of a primordial "RNA World," in which RNA was the original biological molecule that led to all the rest.[121]

Chapter 6

Genes Gone Wild

On the first day of October in 1909, a distraught farmer from Long Island brought one of her prized Plymouth Rock hens to the Rockefeller Institute on Manhattan's Lower East Side. The purpose of the farmer's visit that day was to seek the assistance of scientists at the renowned research institution in a desperate attempt to save her ailing bird.

The chicken was brought to the laboratory of pathologist Peyton Rous, a native of Baltimore and recent graduate of the prestigious Johns Hopkins University School of Medicine. Rous was hired to work in the field of oncology research with the mandate to pursue the research interests of the more experienced Dr. Simon Flexner, the Institute's Director.

At the time, cancer research was not viewed through the same lens as that used in the second half of the 20th century. While some no doubt considered the effort a noble quest, others saw a career in cancer research as a fool's errand, a professional life doomed to the fruitless pursuit of an understanding of an ailment of such complexity that considerable progress would long elude the grasp of human scientific inquiry.

Rous was not amongst those who held this pessimistic view. Neither, evidently, was he interested in setting limits on what he believed might be possible. Although cancer research was in its primitive stages, and no one could have a coherent vision of the potential benefits, or futility, of trying to understand it, Rous planned to apply his medical training as a pathologist to the pursuit of unraveling the secrets of this most mysterious disease.

On that October day, Rous knew he was out of his medical comfort zone trying to administer aid to an ailing chicken. However, since the poor woman was so distraught, he realized he had nothing to lose by looking at the poor bird. Doing so, Rous recognized that the large mass protruding from the chicken's abdomen, sticking out from under the right wing, was a tumor.

He took samples for microscopic examination. Sure enough, he was looking at a *sarcoma*, a potentially malignant cancer found in connective tissues such as the bones, cartilage, muscles, and fatty tissues. This tumor looked like a spindle-cell sarcoma, a rare cancer characterized by the predominance of elongated cells in the microscopic field.

Rous knew the prognosis was grim, and there was nothing he could do for the poor chicken, or for the beleaguered farmer, for that matter. Rous realized the affair was far from a total loss, as it presented an opportunity for interesting (and unexpected) experimentation. He was far too intrigued to let this opportunity pass him by. He knew he had to investigate the case of this dying chicken, which had landed in his lap.

Rous's experiments at the Rockefeller Institute provided two groundbreaking results. First, he was successful in his attempts to transplant samples of the tumor into other chickens of the same brood. In such cases, the transplanted tumors grew rapidly in the genetically matched birds. On the other hand, all attempts to transplant the tumor into chickens that did not share the same familial genetics were unsuccessful.

The explanation for these results was a mystery at the time, as Rous's work on cancer took place when the science of immunology was also in its infancy. Today, we know that just like the rejection of a genetically unmatched, transplanted organ, immune recognition of a tumor in genetically unrelated animals will rapidly destroy the transplanted tumor.

Rous's second finding was a spectacular breakthrough. His 1911 paper in the *Journal of Experimental Medicine*, entitled "A sarcoma of the fowl transmissible by an agent separable from the tumor cells,"[122] described the transmission of the cancer by an extract of the tumor filtered through a membrane known to remove both the chicken cells and bacteria, thereby creating what biologists call a *cell-free extract*. The tumor's transmission by the cell-free extract showed that the tumor could be transmitted to a genetically related bird by an agent smaller than bacteria. Although viruses were the only known infectious agents smaller than bacteria, Rous was circumspect and scientifically exacting, as was his nature. As he lacked definitive evidence, he referred to the source of the infection as an "agent" in his 1911 paper.

Over the next two years, several more infectious agents capable of transmitting cancer in birds were discovered in Rous's lab. Rous was convinced these agents were viruses present in the filtered tumor extract. Unfortunately, he could not think of a way to prove this hypothesis, as the biochemical and immunological tools required for solving the problem would not exist for another half-century.

In the 1970s, a group of American scientists that included Peter Duesberg, Peter Vogt, and future Nobel Prize winners Howard Temin, Harold Varmus, and J. Michael Bishop finally filled in the missing pieces in the story behind the transforming capability of the virus that Peyton Rous had observed early in the twentieth century. Their work demonstrated the presence of certain genes with remarkable properties in the mammalian genome. Under certain conditions, these genes can initiate and/or drive the progression of cancer. They were named *oncogenes*—literally, "cancer-causing genes." The protein products of such genes had the magical power to transform a normal cell into a tumor cell.

The new tools of molecular biology that emerged in the 1970s showed that the extract from Rous's chicken sarcoma did indeed contain a viral gene that directly caused tumor formation in chickens. By studying the action of oncogenes in Rous's virus (later named *Rous sarcoma virus, RSV*) and other tumor viruses, the nature of these strange genes and their relationship to human cancer was elucidated.

In RSV's tiny genome (containing only four genes), there was a gene named *src* (for sarcoma, the type of cancer caused by RSV). The src gene was not essential for viral survival or replication, yet it was responsible for the transmission of the cancer. It was the first of a lengthy list of currently known oncogenes that reside in the genomes of *oncogenic* (cancer-causing) *viruses*.

What was the origin of these magical genes? The answer was found by looking at the genomes of the infected animals. Sure enough, there were animal genes that were about identical to the viral oncogenes, with only slight differences between them, sometimes only a single base change.[123]

When viruses that infect humans were evaluated for the potential presence of oncogenes, the origin story of the aberrant genes also applied for *homo sapiens*. For some of the known human oncogenes, a highly similar gene (called a *homolog*) was found in human retroviruses, the class of virus that includes HIV. This meant that, at least for some human cancers, genes originating in humans can be hijacked by retroviruses and thereby transformed during viral evolution.

The results of the investigation into the origin of oncogenes showed that human genes are hijacked by retroviruses, which infect human (and other mammalian) cells by inserting their tiny genome into the host chromosomes. Like all viruses, retroviruses use the host cell's molecular machinery for their own replication. During the creation of progeny viruses, pieces of human DNA can be extracted from the chromosome along with the viral DNA and packaged into new viral particles.

Viral evolution is rapid, with a far greater rate of mutation (about two to three orders of magnitude) relative to that found in mammalian cells.[124] Once inside the virus, the human DNA can mutate and create damaged forms of human genes that are later inserted back into the human genome during viral infection. Most of this damage does not lead to any noticeable change in phenotype; in fact, the exchange of viral and mammalian DNA is a component of evolution that not only adds genetic diversity, but sometimes actually drives evolutionary change.

Occasionally, however, virally induced damage to a mammalian gene can result in significant pathological effects, including cancer.[125] In the case of oncogenes, further work showed that in their undamaged form, the protein products of these genes participate in critical processes such as the control of cellular proliferation, cellular differentiation, and the regulation of the cell cycle.[126] In their mutated form, the products of such genes can drive tumor initiation and progression.

The human genome is riddled with pieces of viral genes that have been shared over millennia. However, the likelihood that a viral oncogene transferred into humans would result in cancer is low. Given the vast complexity of the human genome, the integration of the viral gene must meet exacting conditions to enable the production of the oncogenic protein. First, the oncogene must be inserted in the correct orientation relative to the transcription machinery's unidirectional movement on the DNA. Next, the insertion point of the gene must be close enough to a promoter to support gene activation; and finally, the oncogene must reside in a reading frame on the DNA that enables efficient transcription.

Only a handful of viruses have been identified that unequivocally participate in the initiation and progression of cancer by the action of an oncogene. For example, there are rare cases of a cancer called Burkitt's lymphoma that are associated with co-infection by malaria and the Epstein-Barr Virus (EBV). EBV, the cause of infectious mononucleosis, is a common human virus that has been associated with several types of inflammatory disease while causing no detectable pathology in most individuals.[127] Recently, intriguing data has emerged suggesting that EBV infection may function as the trigger for multiple sclerosis, causing chronic inflammation that generates autoimmunity.

A few viruses that infect humans cause chronic inflammation that supports cancer progression. Examples include human papillomavirus (HPV) and *hepatitis B* and *C viruses* (*HBV* and *HCV*). In cases where HPV infection leads to cervical cancer or when chronic hepatitis infection leads to liver cancer, the viral infection is not the direct cause of the tumor. Rather, the immune dysregulation that develops in the body under conditions of chronic infection with these viruses establishes a state of smoldering inflammation that destabilizes homeostatic controls over time, thereby driving carcinogenesis.

Along these lines, HPV and the hepatitis B/C viruses do not contain oncogenes that de-regulate cellular processes. Rather, as noted above, they promote an inflammatory environment that, over time, can lead to chronic inflammation with the subsequent development of cancer. Since these viruses do not express a gene that alters cellular phenotypes, they are not formally designated as oncogenic viruses.

The random insertion of a viral oncogene into the mammalian genome in a manner that drives tumor formation is a "needle in a haystack" type of proposition. Still, another means of oncogene activation occurs on occasion. In such instances, an intact human gene mutates while resident in the human genome, changing into the oncogenic form of the protein due to the mutation. Decades of study indicate that the mutation of a normal human gene into an oncogene is a major cancer-promoting mechanism in many mammalian cancers.[128] Such genes are known as *proto-oncogenes*, which we can think of as critical genes with the potential to mutate into active oncogenes.

While this certainly occurs, it is not a frequent event, as a mutation must alter the protein structure in a manner that alters its function. Proteins are tolerant of small biochemical changes. Many point mutations, for example, are silent from a phenotypic standpoint, as they do not alter the three-dimensional shape and/or the binding properties of proteins sufficiently to change their biological activity. Once in a great while, a mutation is introduced that alters the function of a protein critical to controlling cellular functions. Such a mutation might change a growth factor receptor such that, instead of providing a growth signal to the cell only when a growth factor is bound at the receptor, the receptor is permanently locked in the "on" position, thereby providing a constant proliferation signal.

The known proto-oncogenes serve many essential functions in human cells. Some proto-oncogenes are particularly critical during embryogenesis, where they are expressed at elevated levels to help drive rapid cellular proliferation and coordinate the explosive growth and differentiation of the stem cell populations that drive embryogenesis. Later in life, even if unaltered by mutation, a malfunction in genetic regulation can lead to the over-expression of a proto-oncogene, which can send the cell into proliferation overdrive and over the edge into carcinogenesis in the absence of a mutation of a proto-oncogene.

In addition to regulating cellular proliferation and differentiation, proto-oncogenes code for proteins with other important regulatory functions in the cell. Examples include the control of the cell cycle and the regulation of apoptotic signaling. The dysregulation of these critical processes allows cancer cells to separate themselves from the cellular collective, launching themselves on a path of destruction.

In his Nobel Prize acceptance speech in 1966, Peyton Rous said, "Every tumor is made up of cells that have been so singularly changed as no longer to obey the fundamental law whereby the cellular constituents of an organism exist in harmony and act together to maintain it."[129] Just as a free society inevitably crumbles in the absence of the rule of law, cancer cells create molecular chaos as they liberate themselves from the bonds of a fundamental biological edict: the mandate that the cells work cooperatively to maintain homeostasis. No longer subject to the cellular systems of command and control, cancer cells are refractory to the biological signals that maintain the health of the organism.

Cancer cells are so profoundly altered that we might think of them as invaders raining down upon us from another realm, a mad horde of dastardly aliens hell-bent on a path of biological mayhem. The obvious truth is that cancer cells are not intruders at all. They are a part us, spawned by our cells, transformed into cellular monsters that, in Peyton Rous's words, have "somehow been rendered proliferative, rampant, predatory and ungovernable."[130]

As aberrant as these cells may be in their biological behaviors, and as dangerous as they are to the fate of the whole organism, cancer cells do not possess a bag of magic biological tricks. They do not invent novel biochemical pathways, as they cannot conjure from the ether new circuits for control of the pathways they use for their dastardly biological mischief. Rather, cancer cells utilize the same cellular circuits used for maintaining homeostasis, co-opting signaling pathways to alter normal biological processes for their own nefarious purposes.

Cancer is, at its core, a disease of the genome. More precisely, it is a disease caused by the improper functioning of the genome. There are numerous paths to tumorigenesis, including the activation of proto-oncogenes, overexposure to environmental insults (such as solar radiation and reactive oxygen species), damage to tumor suppressor genes, and/or random errors in DNA replication that slip through the cellular quality control system. New cancers arise and can be encouraged to grow by genomic malfunctions, errors in the operation of the source code.

The genetic changes that lead to cancer can result from multiple sources. A small number of cancers, in the range of 1-5%, are inherited cancers.[131] These types of cancers—such as teratomas, as described in Chapter 2—are called *germline cancers*. This is because the embryo—the product of the germ cells, the egg and sperm—is the source of the genetic damage linked to these cancers. Genetic errors in the germ line are passed from generation to generation.

It is important to note that depending on the type of genetic error, and its potential interactions with other genes, the presence of the genetic potential for cancer in these cases does not guarantee that cancer will occur over the lifespan of the affected individual. In fact, it has long been evident that cancers are usually the result of multiple genetic errors. There are several reasons, including the fundamental fact that we bear two copies of each gene. If one is damaged, the other can still operate to achieve the gene's assigned function. Tumor suppressors, for example, remain operational in the cell unless both copies of the gene are corrupted and unable to function properly (an exception is the p53 tumor suppressor, as discussed in Chapter 3).

Since our biological circuitry has inherent redundancy that has developed over our evolutionary history, the corruption of a single biological circuit is unlikely to lead to a failure of the cellular machinery. Given the power of host immunity to root out and destroy nascent tumors and the innate ability of the cellular quality control systems to find and correct errors (or kill the cell if unable to do so), it has long been believed that cancer causation must involve multiple "hits" on the genome.

This concept goes back 70 years. In 1953, British biologist C.O. Nordling advanced the theory that cancer is not the result of one but rather of several mutations.[132] The hypothesis was based on the work of a group of German biologists and physicists, including the renowned physicist Ernest Schrodinger, in the 1940s. The concept hinged on the contention that a single mutational event as cancer's cause would not explain the observed pattern of increased cancer frequency with age.

There was a conceivable argument against the multiple-mutation hypothesis that might support the idea that a single mutational event causes cancer: The latency period of a decade or more between the mutational event and the onset of clinical cancer might be responsible for the increase in cancer incidence with age.[133] The problem with this idea was that it did not provide a sufficient explanation for the relatively high cancer rates observed late in life. A latency period of sixty or seventy years would be required to explain these cases, and this did not seem reasonable.

In a 1943 paper, Albert Dahlberg proposed that malignancy develops with increasing ease with rising age, perhaps because the mutated cells have a chance to multiply over time.[134] This proposal suggested that a single mutation might be all that is needed to promote cancer. This idea did not satisfy some researchers, including Nordling, who proposed that several mutations in the same cell are required and suggested that about seven mutations are needed to cause human cancer.[135]

Amazingly, work performed decades later using statistics on cancer incidence as a function of age demonstrated an exponential sixth-power relationship. Not only are multiple independent events that alter the genome required to produce most cancers (retinoblastoma may be an exception), but this statistical research revealed that the sixth-power relationship provides a mathematical rationale for the contention that the number of genomic events involved in carcinogenesis is about six, perhaps somewhere between five and seven in most cases.

Each cancer is genetically heterogeneous, comprised of multiple subpopulations that are genetically distinct from each other, engaged in a ruthless Darwinian competition for survival. The unique nature of each cancer makes the study of cancer the most challenging medical problem of our age. This also explains why successful treatment of human cancer is so difficult.

In addition to cancers caused by environmentally induced mutations and inherited cancer syndromes, some cancers result from random, sporadic errors in DNA replication and/or mitosis. Such cancers can occur without any of the known risk factors, such as smoking, excessive exposure to toxins, chronic inflammation from viral infection, immunodeficiency, and/or poor nutrition. Such cancers, called *spontaneous cancers*, appear to result from plain old bad luck.

The potential number of DNA replication errors can be estimated. With an error rate of about one base in a billion replicated,[136] and a cellular genome of about three billion base pairs (six billion bases in the double helix), we can expect about six DNA replication errors every time the genome is copied. It is, therefore, accurate to say that uncorrected genetic changes occur every time a cell divides.[137] Given the presence of trillions of cells in our bodies and multiple generations of those cells, year upon year, our DNA is subjected to trillions of replication errors throughout our lifetimes.

While a key question underlying cancer research asks why we get cancer, the significant DNA damage in tumors raises an equally interesting question: Why doesn't everyone get cancer? Given the sheer volume of DNA damage we sustain, how can we avoid the genomic disruption that leads to tumor formation?

We are yet a long way from a comprehensive understanding of the dynamics of tumor formation and growth. However, we know we are built to sustain significant genomic damage, thanks to the safeguards provided by our cellular quality control systems. These systems detect and repair DNA lesions and orchestrate the cell cycle to preclude the entry of cells with damaged genomes into DNA replication and cell division, thereby ensuring the integrity of the genome over future generations.

Since DNA damage is a way of life, and there will always be some damage at any given time in the cellular genome, evolution has provided for a state of homeostatic balance where the quality control machinery tolerates low levels of DNA damage. Higher levels render the cell a target for senescence or programmed cell death. When we consider the cellular "prime directive" to preserve the integrity of the genome, we can recognize the significance of the DNA repair machinery but cannot provide a quantitative estimate of the threshold level of DNA damage that poses a biological threat.

Finally, we are beginning to understand the potential roles of the types of DNA damage present in cancer cells and that cancer is a natural result of the existence of DNA changes that provide the genetic variability required to drive evolution. Yet, there is still a considerable gap in our knowledge about the relationship between DNA damage and cancer and how and when DNA damage leading to cancer occurs throughout the lifespan.

It is particularly challenging to confidently assign causation to the occurrence of specific cancers. In the real world, it is difficult to know about the environmental exposure we experience over a lifetime. Many cancers are likely the result of more than one root cause, a series of genetic changes accompanied by an immune deficiency due to nutritional factors, for example.

An active and ongoing controversy about cancer causation is evident in the scientific literature. The debate was fueled by a 2015 publication in *Science* by Johns Hopkins cancer researchers Cristian Tomasetti and Bert Vogelstein.[138] The latter is a prominent physician-scientist and cancer genomics researcher who has pioneered our understanding of the complex genetics of colon cancer. The authors note the vast differences in cancer incidence for various tissues and organs. On the high end, the lifetime risk of a breast cancer diagnosis (in the U.S.) is 13%, and the lifetime risk of lung cancer is 7%. The lifetime risk of colon cancer is 5%, while, at the low end, the lifetime risk of cancers of the nervous system, including brain cancer, is an order of magnitude lower (0.6%).

The most prevalent cancers occur in organs that undergo a high rate of cellular replacement over the lifespan. These replacement cells are the progeny of the tissue-specific stem cells responsible for repopulating the tissue to compensate for damaged and dead cells. Both the lung and the colon require frequent cellular replacement, as they are exposed directly to the environment and subjected to continuous chemical insults.[139] Conversely, for the organs with longer-lasting cells that only require occasional replacement—the nervous system, for example—the risk is significantly lower. Paradoxically, in the case of the small intestine, an organ exposed to a similar toxic load as the colon, the lifetime risk is also quite low, at 0.2%, for reasons that are not yet understood.

How can these observations be explained? Noting that the source of all the organ's cells is the multipotent tissue-specific stem cells that provide both the differentiated cells in the tissues as well as new generations of stem cells, the authors wondered if there might be a relationship between the number of stem cell divisions in an organ and the incidence of cancer. According to this hypothesis, the more stem cell divisions, the higher the probability of accumulating mutations in the stem cell population.

Using data from the United States for 31 cancer types, the researchers plotted the number of stem cell divisions over the lifetime versus the lifetime cancer risk for that specific type of cancer. While the relationship is imperfect, there is a strong mathematical correlation between the number of lifetime stem cell divisions and cancer incidence. Amazingly, the correlation holds over six orders of magnitude for the number of stem cell divisions, with a low of about a million lifetime divisions for pelvic osteosarcoma (bone cancer of the hip), and a high of about a trillion cell divisions over the lifetime for colorectal cancers. The authors note that "No other environmental or inherited factors are known to be correlated in this way across tumor types."[140]

Tomasetti and Vogelstein published a follow-up paper that extended their analysis beyond the U.S. by examining the relationship between lifetime stem cell divisions and cancer incidence for 17 cancer types in 69 countries. They reported, "The data revealed a strong correlation between cancer incidence and the number of divisions of the tissue-specific stem cells in all countries, regardless of their environment."[141] This global data set demonstrates the same correlations found in the previously published U.S. study. Furthermore, the global data is consistent with the data in America showing that unavoidable random mutations "are responsible for two-thirds of the mutations in human cancers."[142]

While the correlations demonstrated for both U.S. and global cancers are objectively impressive, the papers generated a storm of comment and criticism. One published critique in *Science* noted that "correlation 'explains' the data in the statistical but not the biological sense."[143] This criticism is based on the adage that correlation does not demonstrate causation; in other words, the presence of a statistical relationship between two variables does not definitively prove that one variable exerts a direct influence on the other. This is why proving causation is so difficult.

The determination of cancer causation is a complex business, with many variables to consider.[144] Nevertheless, Tomasetti and Vogelstein's analysis provides an intriguing hypothesis about the potential role in tumor initiation of random errors during DNA replication in stem cells, presumably via the introduction of perturbations to genes that play critical roles in the regulation of cellular processes.

The concept that bad luck may play a significant role in cancer causation suggests that many cancers may not be preventable, "that not all cancers can be prevented by avoiding environmental risks."[145] This unsettling thought is countered by the contention in a 2017 *Science* article from Tomasetti, Li, and Vogelstein that the recognition of the contribution of these random mutations "does not diminish the importance of primary prevention."[146]

The previous considerations suggest that many cancers are unavoidable. This line of argument is consistent with the idea that cancer is an innate property of biological systems resulting from imperfections that are, in turn, required to provide the genetic variability needed to drive evolutionary change.

The causal connection between mutations and chromosomal anomalies with cancer occurrence has been known for a century. Only recently have we realized the extent of the daunting complexity of the genetic landscape of cancer cells. By studying the genetic damage in cancer cells and by considering the chaos of the cancer genome, we can find clues to the origin of tumor formation and progression.

According to the *somatic mutation theory of cancer*, a succession of mutational events provides the root cause of cancer initiation and progression. These genetic alterations, which impact genes essential for the maintenance of the genome's integrity and the control of the cell cycle, are hypothesized to occur amidst the enormously dynamic nature of the genome.

The mutational events that lead to cancer occur in conjunction with the considerable number of genetic alterations that occur naturally, with each cell suffering an estimated 20,000 or more daily DNA damaging events from environmental causes, with 10,000 or more DNA replication errors per day for proliferating cells.[147] The responsibility for cleaning up the genetic detritus and preventing the replication of damaged DNA belongs to the DNA Damage Response and the molecular circuitry responsible for regulating the cell cycle. Given the extent of routine genomic damage it is obvious that the genome's innate quality control systems are highly robust, protecting against the relentless attacks on the integrity of our genes.

The somatic mutation theory is consistent with the idea that tumor development takes about six accumulated genetic hits that impact the ability of our cells to preserve the integrity of our DNA and properly direct each cell through the cell cycle. Such mutations can lead to cancer if they provide a growth advantage to the affected cell. A mutation that is directly linked to cancer initiation and progression is called a *driver mutation*. Examples include the mutation of a proto-oncogene to an active oncogene, the loss of tumor suppressor function, the inactivation of caretaker functions that protect the integrity of the DNA, and the loss of gatekeeper functions that regulate the cell cycle.

A DNA mutation that does not provide a growth advantage is called a *passenger mutation*, implying that it does not play a role in tumorigenesis or tumor progression. As noted above, the evidence suggests that about six driver mutations are required to initiate a tumor in most cases, with many accompanying passenger mutations present in most cancer cells.

Given the impact of environmental and internal sources of biochemical damage to the DNA, a damaged cell might be directed by the quality control systems into a senescent state, a state of metabolic stasis designed to give the repair machinery one last chance to make the required corrections to the DNA before cell division. If repair is unsuccessful, such a cell, under homeostatic conditions, receives biochemical signals that trigger apoptosis.

The advent of high throughput genetic sequencing has opened opportunities in the study of cancer genomics. A project called the Cancer Genome Atlas, sponsored by the National Institutes of Health and the National Cancer Institute, has documented the sequences of the protein-coding genes for thousands of cancers, both primary tumors as well as metastatic lesions, across a broad array of cancer types.[148] The sheer volume of mutational errors, the vast heterogeneity within cells from a single tumor mass, and the mutational individuality of each cancer show that each tumor is unique at the level of the genome, with anywhere from dozens to thousands of mutations.[149] In addition, since tumors evolve over time, any mutational analysis performed is a snapshot in time of a dynamic process that renders continuous change in the mutational landscape.

The data derived from the cancer incidence-age correlation analysis suggesting that six or seven mutations were required explains not only the observed incidence-age correlation, but it is also consistent with the data on cancer driver mutations. While the detection of genetic damage in cancer cells was anticipated, the extensive mutational diversity observed in cancer cells was surprising. The most striking finding was the presence of thousands of point mutations in some cancers, thereby raising a confounding question: how can these cells continue to operate when there is so much genetic damage?

Many point mutations have minimal impact on protein function. Even in cases where the protein is impacted, many proteins are not critical for maintaining homeostasis. Therefore, many (likely most) of the point mutations in cancer cells do not impact the cell's growth and survival in a significant way. Those changes that are physiologically significant, the driver mutations, often exist amidst a sea of passenger mutations that are generated during cancer development and progression.

Given the fidelity of DNA replication—cited in a range from one error per 100 million to one error per billion base pairs in normal eukaryotic cells—the presence of so many mutations creates a dilemma: how can cells generate so many point mutations? A 1974 proposal by Lawrence Loeb of the University of California at San Francisco was presented to resolve this quandary: "I put forth the concept that the mutation rate of nonmalignant cells is insufficient to generate the large numbers of mutations in cancer cells." He proposed that cancer cells have significantly higher mutation rates relative to normal cells and called this defining characteristic the *mutator phenotype*.[150]

In the five decades since Loeb's concept was proposed, a large body of evidence has accumulated supporting the core idea of the mutator phenotype hypothesis. Most convincingly, genomic data, such as that found in the Cancer Genome Atlas, shows that thousands of mutated genes are often found in cancer cells. Some tumors have a staggering number of mutations, ranging from hundreds to thousands. Sometimes, individual genes have multiple mutations. Some genes for critical proteins, such as the p53 tumor suppressor gene, can undergo dozens of different mutations in cancer cells. However, there is no common set of mutated genes found in all samples from specific types of cancer.

At its mechanistic heart, the increased mutation rate in cancer cells is due to the loss of caretaker and gatekeeper functions. The former function includes the protein networks that effectively detect and repair DNA damage. The latter includes the proteins that regulate the cell cycle in a manner that ensures the genetic integrity of future generations.

With the loss of critical caretaker and gatekeeper functions, cancer cells develop chromosomal abnormalities of various kinds, from the numerous point mutations scattered throughout the genome to other types of chromosomal damage. These genetic changes in cells are passed on in surviving cancer cells to the next generation to create a genomic composition that deviates increasingly, over time, from that found in normal cells. As cancers grow, both the extent and rate of genetic damage continue to increase, thereby rendering the acceleration of genetic instability characteristic of tumor progression.

By the late nineteenth century, biologists knew that cancer cells have observable alterations in the appearance of their chromosomes under microscopic examination, with nuclei showing the presence of multiple types of chromosomal abnormalities. These abnormalities include duplications of sections of chromosomes that are either pasted together on the chromosome of origin or inserted in a different chromosome. There are chromosomes where sections have been ripped away, a phenomenon called *deletion*. Sometimes, large segments of DNA are lopped off the parent chromosome and found either standing alone or attached to another chromosome, a phenomenon called *translocation*.

The most renowned example of such a translocation, known as the *Philadelphia Chromosome*, was discovered in 1960 at the University of Pennsylvania in cells from patients with *chronic myelogenous leukemia (CML)*. In CML, the transformation of a hematopoietic stem cell to a cancer cell involves the translocation of a segment of chromosome 9 to chromosome 22.

This translocation generates a new protein from the fusion of two genes, one on each chromosomal piece involved in the translocation. The fusion generates a protein called BCR-ABL, a mutated form of ABL1 tyrosine kinase. ABL1 is a signaling protein that adds a phosphate group to the amino acid tyrosine on another protein, also a tyrosine kinase, which is activated by this process (called *phosphorylation*) via a conformational change induced by the addition of the phosphate to the protein. Upon phosphorylation, the activated tyrosine kinase similarly activates another kinase enzyme in a signaling cascade that leads to cell proliferation. This structural alteration of the mutated BCR-ABL kinase locks the kinase in the "on" position, such that a continuous proliferation signal is elicited, driving the cell into uncontrolled cell division.[151]

This landmark discovery by David Hungerford and Peter Nowell brought a new appreciation for the role of genetic abnormalities in cancer. Three decades later, this pivotal discovery provided the foundational science for the first successful targeted cancer drug, Gleevec. Approved in 1998, Gleevec blocks the activity of the BCR-ABL fusion tyrosine kinase, thereby turning off the continuous proliferation signal.[152] The discovery of the Philadelphia Chromosome demonstrated a direct link between a specific genetic change and cancer development. The translocation event that moved together the ABL and BCR proteins—juxtaposed in a manner that created a hybrid ABL kinase lacking an "off-switch"—provided conclusive proof for the role of a chromosomal aberration in driving cancer initiation and progression.

In the case of Philadelphia Chromosome-positive (Ph$^+$) CML (about 95% of CML cases), a chromosomal translocation creates an oncogene, not by activating a proto-oncogene to its tumor-promoting form via a mutation in the DNA sequence but, rather, by fusing the proto-oncogene with another protein to create a hybrid oncogene.[153]

The chromosomal translocation event responsible for Ph$^+$ CML was discovered a quarter century before the invention of the polymerase chain reaction that provided a way to analyze genetic changes at the level of the DNA sequence. While there was no means for characterizing mutations in DNA in 1960, the phenomenon could be observed upon microscopic examination of the chromosomes (Figure 11).

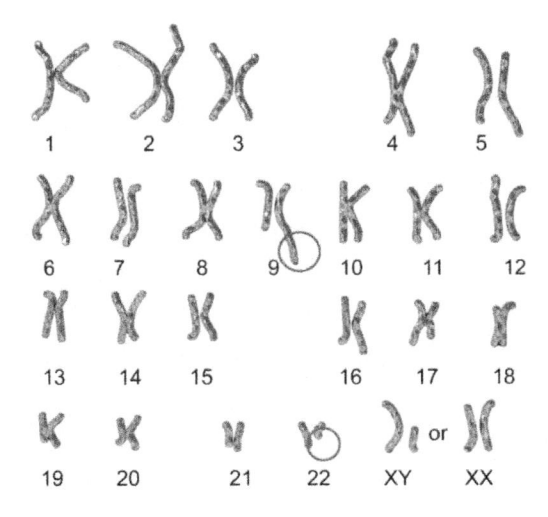

Figure 11. The Philadelphia Chromosome

Note that the section of chromosome 22 that is missing can be found attached to the end of one copy of chromosome 9. The alteration of chromosome 9 creates an oncogene that drives cellular proliferation. There are 23 chromosome pairs in humans, with 22 somatic chromosomes and a single pair of sex chromosomes (XY or XX for males or females, respectively) as shown in the bottom right of the figure.

Right on a microscope slide, Hungerford and Nowell had shown the genetic basis of a type of human cancer. This breakthrough discovery raised the intriguing possibility that other chromosomal rearrangements might be diagnostic for certain cancers. In the best-case scenario, it even seemed possible that all human cancers could be categorized by accompanying chromosomal alterations.

Over time, it became clear that while a handful of chromosomal aberrations could be associated with certain cancers, most cancers lack any diagnostic features based on microscopic evaluation of the chromosomes (the chromosomal pattern upon visual examination is called a *karyotype*). Unfortunately, most cancer karyotypes are unhelpful in providing diagnostic and/or therapeutic clues, since the chromosomal complement of cancer cells is, in general, far more chaotic than the indisputable chromosomal rearrangement observed by Hungerford and Nowell.

Even in normal cells, the chromosomes may not be biochemically flawless. Minor aberrations, including translocations, replications, deletions, and dozens of point mutations, are commonly found. In addition, normal cells can undergo changes leading to the replication or deletion of entire chromosomes, sometimes as a response to environmental stress. For example, hepatocytes (liver cells), under the stress of conditions requiring the detoxification of a significant load of cellular toxins, can replicate chromosomes to produce higher amounts of the protein machinery involved in the detoxification process to maintain homeostatic balance. The flexibility of our genetic programming allows for chromosomal changes in response to stress.

Under normal conditions, our cells contain two copies of each chromosome, a condition known as *diploidy* (the adjective is *diploid*). A cell containing more than two copies of each chromosome is called a *polyploid* cell. A gamete, which has only one copy of each chromosome, is called a *haploid* cell (from the Greek word *haplous*, meaning "single"). A cell containing a deviation from the diploid chromosomal number—for example, one extra chromosome or one missing chromosome—is called an *aneuploid* cell (such cells are in a condition called *aneuploidy*).

German zoologist Theodore Boveri's work with sea urchin embryos in the late nineteenth century demonstrated that the presence of aneuploidy has profound biological effects. Using an experimental system that allowed for the generation of abnormal mitotic divisions resulting in three or four cells instead of the customary two, he created aneuploid cells and observed the impact of their chromosomal abnormalities on cell physiology. The results demonstrated that aneuploidy is a highly detrimental and often lethal biological condition. Based on his work and observations made by others, Boveri went on to propose that aneuploidy may play a role in cancer.[154]

A century later, we know that aneuploidy is present in many human cancers. The most critical question in this regard is whether aneuploidy is a cause of cancer initiation and progression or a consequence of the genetic instability that characterizes cancer. From a biological standpoint, the critical evaluation of this question starts with the recognition that the presence of an extra chromosome (and/or the absence of one chromosome from the typical pair) can directly impact the levels of the protein gene products synthesized by the cell.

This consideration is embodied in the concept of *gene dosage*. Let's say that chromosome 6 contains one copy of gene X. In a diploid cell, we would anticipate the production of gene product X (i.e., protein X) at about twice the level that is found in a cell with only one intact copy of gene X, such as a diploid cell that has lost one functional copy of gene X due to mutation, or an aneuploid cell with only one copy of chromosome 6.

In the case of aneuploidy where an extra chromosome is present (for example, three copies of chromosome 6), there are three copies of gene X, leading to the production of approximately three times the level of gene product X relative to a cell containing only one functional copy of gene X. We can say that the cell with three copies of chromosome 6 (and three intact copies of gene X) has three times the gene dosage relative to a cell with only one intact copy of gene X.[155]

Several mechanisms are in play in cancer cells that alter gene dosage. A gene (a tumor suppressor gene, for example) may be mutated to an inactive state. Alternatively, the region of the genome containing the gene might be lost, subjected to a deletion event that removed a gene or section thereof. Conversely, the gene dosage can be increased, for example, by gene duplication or the activation of a normally quiescent gene by its translocation to a region impacted by an active promoter or enhancer. Aneuploidy has the potential to either increase or decrease gene dosage, depending on whether the aneuploidy is due to the duplication or deletion of genetic material. Such changes in gene dosage can have profound effects, including the development of cancer (for example, by over-expression of a proto-oncogene).

Leonard Hayflick, a young scientist at the Wistar Institute in Philadelphia in 1961, was tasked with growing several types of cells in culture for the more experienced researchers at the prestigious biomedical research institute that was founded in 1892. While cell culture techniques have been routinely used in research and pharmaceutical/biotechnological product development since the mid-seventies, successfully growing cells under laboratory conditions in 1961 required a unique skill set and experience base not widely available.

The cell lines used for cancer research at the time, however, grew easily and sometimes indefinitely in culture. After all, cancer cells have the uncanny ability to "never say die" unless confronted with therapeutic interventions. Even then, there is no guarantee they will be forever stymied in their obsessive drive to divide and multiply.

However, Hayflick wasn't going to have an assignment as easy as growing cancer cells. He was asked to grow normal human *fibroblasts*, cells found on epithelial surfaces that can be easily obtained by swabbing inside the cheeks. When he grew non-cancerous (also known as *non-transformed*) cells, he observed that they stopped dividing after anywhere from about 50 to about 80 rounds of cell division. After undergoing these divisions, the cells remaining alive were senescent, characterized by their inability to divide again. This interesting finding, dubbed the *Hayflick Limit*, demonstrated that our cells are programmed to stop dividing at a defined point during their lifespan. Hayflick's work showed that cells have a built-in barrier that limits the duration of their proliferative capacity.

While non-transformed cells have a finite proliferative lifespan, cancer cells can blow right by the Hayflick Limit when grown in cell culture. The basis of this intriguing finding would not be biochemically elucidated for three decades, as the analytical tools were not available to unravel this mystery in 1961.

The secret of the Hayflick Limit began to reveal itself in 1990, when

C.B. Hurley of McMaster University (Ontario, Canada) observed that the chromosomes shorten ever-so-slightly with each cell division. The biochemical basis of this phenomenon was revealed by the discovery of a special structure at the ends of chromosomes called *telomeres*.

The telomeres are comprised of DNA complexed with specific proteins. The telomeric structures contain a repeated six nucleotide (*hexanucleotide*) sequence, TTAGGG (on one of the strands of the double helix, with the complementary sequence AATCCC on the other), repeated over a stretch several kilobases (thousands of bases) long. At the end of the chromosome, one of the double strands of the telomere ends prematurely, such that a single-stranded "overhang" of the telomere (several hundred bases long) loops back toward the double helix.

This looped structure, called a T1 loop, is bound to a protein complex called *shelterin*. Together they form a "knot" at the end of the chromosome to prevent aberrant DNA repair. The presence of a "blunt end" would trigger the DNA Damage Response that maintains the integrity of the genome. The abrupt cessation of the double helix in a blunt end would be indistinguishable from a double-stranded DNA break, which poses a significant threat to cellular survival.

The telomere also contains an RNA sequence that serves as the template for DNA synthesis on one of the two strands of the DNA. Such RNA primers are required for DNA replication on one of the two strands because DNA polymerases and the accompanying DNA replication machinery can only move in one direction along the DNA template.

The presence of this protein-nucleic acid complex (containing DNA and RNA) at the end of the chromosomes has an impact on the ability of the DNA replication machinery to copy the DNA at the chromosomal ends (the telomere's hexanucleotide repeats). Thus, with every DNA replication cycle, DNA replication at the last few nucleotides is not possible due to the presence of the telomere, which blocks the DNA replication machinery from having access to the end of the DNA sequence. This causes the replication machinery to lose its "biochemical grip" on the DNA at the end of the chromosome.

This leads to a shortening of the chromosome's end, where the number of telomere repeats reduces after each cell division. This biochemical mechanism provides a "biological clock" that regulates the cell's lifespan. From an evolutionary standpoint, this provides a checkpoint for older cells, which contain unrepaired, damaged macromolecules that increasingly accumulate. This biochemical restriction, a fail-safe mechanism to guard the organism against the accumulation of damaged (and potentially tumorigenic) populations of cells, gradually shortens the chromosome ends until they reach the point where the telomeres are sufficiently shortened to preclude the faithful replication of the DNA strands. This results in the arrest of the cell cycle and the triggering of apoptotic signaling that sounds the death knell of the individual cell for the sake of the organism's survival.

Further investigation revealed that the ever-ready tumor suppressor p53 acts as a biochemical sensor for shortened telomeres, activating the program of cellular senescence in cells with telomeres that fall short of the minimal length required for division. It has been noted that the telomeres protect the chromosomes from damage while also serving as an early warning system for damage to DNA integrity—a "'canary in the coal mine,' alerting the cell's quality sensors before damage to the genetic coding sequences occurs."[156]

In the absence of active p53 protein, the cell can continue replicating despite its shortened telomeres. This accelerates chromosomal instability and the creation of a cellular crisis. In the best case, this results in apoptosis; however, if the apoptosis machinery is damaged (such as by the loss of both the p53 and Rb proteins), a cancer cell may arise due to the loss of functional p53.

An enzyme called *telomerase* is responsible for regulating telomere length. While telomerase can extend the length of the telomeres, it is present at low levels in normal cells, and at higher levels in cells that can replicate beyond the Hayflick limit. Stem cells express telomerase in appreciable amounts, as do cancer cells.

Current data suggests that 80-90% of cancer cells express the telomerase enzyme in amounts significantly higher than those found in normal human somatic cells.[157] While telomerase over-expression is present in most cancer cells, whether this is a cause, or an effect, of cellular transformation remains unclear. Over-expression in and of itself is not tumorigenic in the presence of intact p53.

Recent evidence shows that telomere shortening is present during the initial stages of tumorigenesis and that the telomeres appear to lengthen later in cancer development. Analysis of protein expression indicates the presence of appreciable levels of telomerase during the later phases of cancer progression, thereby allowing the cancer cells to continue dividing. In addition, the lengthened telomeres resulting from telomerase expression protect the cancer cell from excessive levels of chromosomal instability that can cripple the cell's replicative machinery, as telomere dysfunction can increase chromosomal instability.[158]

By expressing elevated levels of telomerase, the cancer cell ensures the continuation of DNA replication and thereby heightens the cancer cell's chances of replicating its aberrant genome for the genesis of future generations of cancer cells.

Divide, divide, divide! The cancer cell wouldn't have it any other way.

Chapter 7

Chaos by Design

The discovery of the Philadelphia Chromosome—which showed that a genetic translocation can drive tumorigenesis—led to an intensive search for other genetic defects in cancer beyond the previously identified mechanisms of proto-oncogene activation and tumor suppressor loss. The results of this research provide a deeper understanding of cancer that encompasses not only damage to individual genes, but also the changes in chromosomal composition found in cancer cells.

Recent evidence substantiates the importance in cancer of the genetic phenomena described in the preceding chapter, including the mutator phenotype, aneuploidy, replications, deletions, and translocations. These phenomena are embodied in the modern concept of genetic instability as a fundamental property of cancer cells, characterized by continuous alterations in the genome's structure and content. The profile of genetic instability in cancer includes the presence of small fragments of chromosomes in either linear or (more commonly) circular form. The phenomenon that leads to the presence of these chromosomal fragments is called *chromosomal shattering*, or *chromothripsis* (*thripsis* means "pressure" or "distress" in Greek).

Genetic instability is now recognized as a common feature in cancer cells. For most of the cancer cells, these ongoing changes inevitably lead to the cell's destruction, as the accumulated genetic damage eventually precludes successful DNA replication and cell division. The tumor cells that survive this arduous evolutionary process, those best adapted to the tumor's *microenvironment* (that is, the local conditions in the tissues surrounding the tumor), provide the "seeds" that drive tumor progression.

Over the past decade, a prolific body of data on the genetic alterations found in human cancer has accumulated, posing the significant challenge of understanding the relationship between these various forms of genetic damage and cancer initiation and progression. The development of such an understanding would provide important clues to the mechanisms underlying the cancer cell's natural history.

The presence of multiple forms of chromosomal damage creates a multifaceted "chicken and egg" question: which came first, the aneuploidy or the multitude of point mutations? Where do deletions and replications of genes and chromosomes fit in during tumorigenesis and metastasis? The staggering biological complexity of these questions goes right to the heart of our understanding of cancer. While cancer is clearly a genetic disease, the genome's characteristics, and how chaotic cancer genomes function during the various stages of human cancer, remain enigmatic.

The genomes of cancer cells are highly atypical. Most human cancers, 90% or more, contain aneuploid cells. Studies in yeast—a eukaryotic organism that, while distant from us on the phylogenetic tree, has biochemical systems that are highly similar to our own—demonstrated that the presence of aneuploidy alone can drive genetic instability. There is evidence in certain yeast strains that the presence of "2N+1" aneuploidy (a full set of chromosomes plus one extra chromosome) can directly increase the rate of point mutations. These results show that, in some circumstances, aneuploidy can drive the creation of the mutator phenotype found in cancer cells.

The segregation of the chromosome sets during cell division is a critical process that has its own quality control checkpoint called the *mitotic checkpoint*. The biochemical machinery (proteins) responsible for this checkpoint have an essential function in the production of new cells. These proteins ensure each cell contains the appropriate chromosomal complement after cell division. If not, the daughter cells are recycled to create the building blocks (amino acids, nucleotides, sugars, and lipids) for the biosynthesis of new macromolecules.

Occasional failures at this checkpoint enable the erroneous distribution of the chromosomes, a process called *missegregation,* which results in the distribution of one (or more) chromosomes to the wrong side of the cell before cytokinesis. In a case where one chromosome is wrongly distributed, both daughter cells are aneuploid, with one containing a full diploid set of chromosomes plus one extra chromosome (a 2N + 1 cell), while the other daughter cell is missing one chromosome from the full 2N diploid set (a 2N-1 cell).

Aneuploidy can contribute to further deterioration of chromosomal stability, enhancing the rate of biochemical errors that result in additional chromosomal rearrangements, deletions, and replications. The cell can enter a state of increasing aneuploidy that contributes to further chromosomal instability.[159] Alternatively, aneuploid cells can continue to produce aneuploid daughter cells with the same defect in the chromosomal complement as the parent cell—a case in which the aneuploidy characteristic is stable, even if the overall genetic composition is not.[160]

The tremendous diversity of chromosomal aberrations in cancer cells provides the extensive genetic variation required for the evolution of increasingly aggressive populations of genetically related cells (clones). In this manner, chromosomal instability "is the driver of cancer progression."[161] While most of the genetic aberrations are either non-impactful or detrimental to cellular growth and survival, there are, on occasion, genetic aberrations that provide a competitive growth and survival advantage for a particular cancer cell clone.

A successful cancer clone—one that has out-competed its rivals—is likely to accumulate other biochemical defects along the way, sacrificing homeostasis to enable the prioritization of cellular adaptation to the unrelenting stresses of the competition: "There is little doubt that continuous genetic change is one of the key factors underlying the adaptive evolution of malignant cells."[162] This hallmark characteristic of cancer cells significantly complicates treatment, as it "makes cancer a 'moving target' for defensive responses of the host or the attacks of anticancer treatments."[163]

Given cancer's biological complexity, formulating a coherent and comprehensive description of how cancer cells operate and impact the health of the afflicted individual is a formidable task.

While it is critical to discuss the cancer cell's characteristics and behaviors, it is important to consider that cancer is a systemic disease that impacts many of the body's physiological processes. As tumors form and grow, they interact with normal cells, both in the local region of the tumor as well as throughout the body. Therefore, the biology of cancer can be divided into a description of how tumor cells perform their functions (in this chapter), followed by descriptions (in subsequent chapters) of how cancer cells interact with their environment, including how healthy cells are co-opted by cancer cells to help drive tumor progression.

One of the premier challenges of modern cancer research rests in understanding how the genetic changes found in cancer cells drive tumor formation and progression. The identification of these *driver genes* provides a framework for developing such an understanding. As noted previously, current data shows that anywhere from two to seven driver mutations are sufficient to launch a cell on the path to tumor formation and progression. The driver genes are altered forms of tumor suppressors and proto-oncogenes that, in their unmutated state, suppress tumor formation and regulate key cellular processes, respectively.

While about 200 driver genes have been identified in human cancers, these genes impact only about a dozen signaling pathways. The caretaker and gatekeeper proteins control these pathways, ensuring DNA integrity and controlling cellular proliferation, differentiation, and survival.

Analysis of samples from various cancers demonstrates a broad range of mutations in the different cancer types. Solid tumors—such as breast, brain, and pancreatic cancers—contain dozens of mutations. Genomes from melanoma and lung cancer (in smokers) have hundreds, or even thousands, of mutations, presumably because of exposure to ultraviolet light (in melanoma) and the various toxins in cigarette smoke (in lung cancer). In the rare cases of lung cancer in non-smokers, the number of mutations found is about tenfold less than that found in lung cancer patients with a history of cigarette smoking. The status of the DNA repair pathways has a significant impact on the mutational burden found in human cancers, as tumors in patients with DNA repair defects often harbor thousands of mutations.[164]

In addition to the observed genetic changes, alterations in gene expression through epigenetic modifications to the histone proteins and the DNA are common in human cancers. It is difficult to determine if specific epigenetic alterations are causative in driving cancer or merely the natural outcome of the growing tumor's presence. We can conceptually differentiate an epigenetic driver from an epigenetic passenger, though in practice it is not yet feasible to discern the differences between the two types of epigenetic changes. Achieving a detailed understanding of the significance of epigenetic alterations in cancer, which occur in an unstable and dynamic genome, presents one of the most daunting challenges in modern cancer research.

The proliferation of normal cells is tightly regulated to ensure that the proper numbers and types of cells are present in the various tissues throughout the lifespan. The regulation of the rate of cellular reproduction is controlled by two opposing classes of signaling molecules: those that stimulate cell division and those that shut it down.

Proliferation signals are local—that is, they only influence cells in the region where the signals are generated—and the biological effect of the signaling is short-lived. Under homeostatic conditions, cells will not undergo mitosis unless they receive the molecular signals that stimulate the passage of a cell through the cell cycle into DNA synthesis and mitosis. Following proliferative signaling, the interaction between a stimulating molecule, such as a growth factor, and a signaling protein anchored in the cell membrane, such as a growth factor receptor, rapidly degrades. In this manner, tight control is achieved over the critical processes that govern cellular reproduction in response to local conditions.

Receptors are complex and fascinating proteins. They are members of a class of proteins that connect the inside of the cell with its exterior. These proteins, which traverse the cell membrane, are (for obvious reasons) called *membrane proteins*. Membrane proteins contain three functional parts. The extracellular domain protrudes from the cell surface into the spaces between cells, where it binds specifically to a protein partner known as its *ligand*. The transmembrane domain spans the cell membrane (sometimes more than once), connecting the extracellular domain outside of the cell with an intracellular domain inside the cell.

The binding of the ligand by the extracellular domain alters its three-dimensional shape such that it sends an activation signal (through the transmembrane domain) to the intracellular domain. The activated intracellular domain interacts with other proteins that set off signaling cascades involving proteins and small molecules (e.g., lipids, nucleotides, and ions) that elicit a biological function, such as stimulating the expression of a gene (or set of genes).

For example, after a growth factor receptor binds its ligand (a growth factor), a protein kinase is activated by the receptor's intracellular domain when the kinase adds a small chemical group (phosphate) to another protein. This targeted chemical modification, called *phosphorylation*, induces a structural change that activates the target protein. The target protein might be another protein kinase, which, upon activation, adds phosphate to another protein kinase, and so on, sometimes several protein kinases deep, until the signal is passed on to other molecules for further processing by the cell. Such an arrangement is called a *protein kinase cascade*.[165]

Protein phosphorylation is ubiquitous in nature, found in everything from bacteria to humans. It is an extraordinarily successful evolutionary strategy for effective protein signaling. Numerous oncogenes have been identified as aberrant forms of protein kinases involved in the regulation of the cell cycle.

The ability to replicate without requiring signaling from the external environment is one of the key properties contributing to the proliferative success of cancer cells In the case of cancer cells, which liberate themselves from the constraints of homeostasis, By freeing themselves from this limitation, cancer cells can replicate at rates that far exceed those seen at any time in the life cycle, with the possible exception of some of the accelerated growth processes of embryogenesis. There is, of course, an enormous and crucial difference between embryogenesis and tumor growth. While rapid cellular proliferation in embryogenesis is a planned event programmed by the genome to drive the growth and differentiation of the embryo, cancer cells can achieve growth rates that are both uncontrolled and undesirable.

Cancer cells use several strategies to bypass the requirement for exogenous growth factor signaling prior to cell division. The cancer cell can increase growth factor signaling by overpopulating the cell surface with growth factor receptors and/or by producing mutant growth factor receptors with a permanent "on switch" that provides a continuous proliferation signal that is independent of growth factor stimulation. The latter phenomenon, which drives uncontrolled cellular proliferation, is often observed in cancers of the breast, colon, lung, stomach, and brain in the form of mutations in a growth factor receptor called the *epidermal growth factor receptor (EGFR)*.[166]

There are other molecular tricks the cancer cell uses to create a state of self-perpetuating proliferation. Mutant proteins can sometimes mimic growth factors or alter the intracellular signaling cascade's responsiveness downstream from the receptor's primary signaling interaction elicited by ligand binding at the cell surface, thereby rendering the primary signal superfluous. The cell might produce elevated levels of endogenous growth factors by increasing the expression of growth-factor genes—a process known as *genetic upregulation*—by means of gene duplications and/or epigenetic modifications that increase the expression of these genes.

Regardless of the mechanism involved, the cancer cell strives to become independent of the need for signaling from external sources to break free from homeostatic controls: "This liberation from dependence on exogenously derived signals disrupts a critically important homeostatic mechanism that normally operates to ensure a proper behavior of the various cell types within a tissue."[167]

The evolving cancer cell also develops the ability to circumvent the anti-growth signaling used in normal cells to balance the biological effect of growth-factor pathways. An array of proteins that interact with growth factors and their receptors are at play to contain cellular proliferation within the appropriate tissue-specific ranges for maintaining homeostasis. Such a system allows for adjustments in the rate of cellular proliferation in response to rapidly changing conditions, such as at times of increased cellular stress.

The anti-growth signaling proteins serve an important gatekeeper function by regulating the cell cycle. They function as cellular sensors responsive to stresses induced by environmental conditions that threaten cellular survival. Since these proteins prevent the aberrant replication of cells, thereby guarding against the likelihood of genomic damage in future generations, they are tumor suppressors.

The p53 protein is the quintessential example of an anti-growth tumor suppressor, as its activity can suspend the progression of the cell cycle in the presence of extreme stresses (such as genomic damage and/or shortages of critical biochemical components).[168] The reduced functionality of the cellular circuits that inhibit cellular replication observed in many cancers (such as the loss of functional p53 and/or retinoblastoma protein) enables the cancer cell to become refractory to anti-proliferation signals. Thus, the loss of tumor suppressor function provides a proliferation advantage over cells subject to the homeostatic constraints that carefully regulate the growth and death rates of cellular populations.

The "auto-destruct" process of apoptosis provides, under catastrophic conditions, the final tumor-suppressing barrier that ensures a damaged cell with a corrupted genome is eliminated from the organism. The apoptotic machinery is comprised of protein sensors that detect signs of stress-induced cellular damage, as well as other proteins responsible for the implementation of the auto-destruct program. These sensors monitor molecular stress signals sent from inside and outside the cell, reflective of the condition of the tissue of which the cell is but a part.

Once the apoptotic program is initiated, a series of biochemical reactions are triggered by the effector proteins of the apoptotic cascade, resulting in the molecular breakdown of the cell membrane, the cell's cytoplasmic structures, and the components of the nucleus, including the chromosomal material. This process of molecular destruction, which involves the p53 protein as a key sensor and activator molecule, occurs over a span of 30-120 minutes. The cellular breakdown products are recycled as building blocks for the construction of the macromolecular components of new cells.[169] The critical importance of the p53 tumor suppressor protein in maintaining homeostasis is evident in the cancer data, as about half of all human cancer genomes carry at least one damaged p53 gene.

As tumors evolve, the tumor cells' sensitivity to apoptotic signaling diminishes over time, as the various pathways responsible for apoptosis are corrupted through the faulty molecular circuitry rendered by cancer's evolution. "The multiplicity of apoptosis-avoiding mechanisms presumably reflects the diversity of apoptosis-inducing signals that cancer cell populations encounter during their evolution to the malignant state."[170]

Under homeostatic conditions, the biological components of our cells are subject to molecular quality controls that ensure the repair or replacement of damaged macromolecules. Because damage inflicted upon our proteins by reactive oxygen species and other unstable metabolic byproducts can impair their function, an elaborate quality control system (comprised of multiple sets of binding proteins and enzymes) is ever vigilant in the critical task of sensing, repairing (if feasible), or destroying (if necessary) proteins that have been damaged while performing their molecular duties.

Under stressful conditions—such as nutrient depletion (starvation), chronic illness, or other causes—a cell can recycle damaged organelles to create macromolecular constituents for new, healthy organelles. To achieve this, the cell activates a biochemical process called *autophagy*—literally, "self-eating"—in which damaged cellular constituents are degraded by enzymes in lysosomes.

These enzymes break down the macromolecular components of the organelle into their molecular building blocks, which are subsequently used for the biosynthesis of new macromolecules. In this way, autophagy prevents the build-up of damaged and defective cells and macromolecules. Recent data show that autophagy is critical to the development of the embryo, cell differentiation, and host immunity.[171] Defects in autophagy have been linked to several serious diseases beyond cancer, including diabetes and neuro-degenerative disorders such as Huntington's disease, Alzheimer's disease, and Parkinson's disease.

The 2016 Nobel Prize in Physiology or Medicine was awarded to Japanese cell biologist and geneticist Yoshinori Ohsumi for his pioneering work in the discovery of autophagy, a previously undisclosed cellular mechanism used to maintain the balance of energy and macromolecular raw materials. The discovery's significance was underscored by the rare occurrence of a single recipient of the Medicine prize, which is usually shared amongst the researchers who pioneered a significant branch of biomedical research. In this case, a lone scientist drove this achievement all on his own.

Born in Fukuoka, Japan, in February 1948, Ohsumi noted his humble roots in his Nobel Prize Banquet Speech. Trained at the University of Tokyo and the Rockefeller University (formerly the Rockefeller Institute in New York City), he considered himself a "basic cell biologist who has been working with yeast for almost forty years."[172] He first observed autophagy in the microscope in 1990 while working as an assistant professor at the

Tokyo Institute of Technology. He identified 15 genes that code for proteins involved in the autophagy pathway.

Like Sydney Brenner before him, he acknowledged the critical role that his model organism, *Saccharomyces cerevisiae*, also known as common brewer's yeast, played in his voyage of scientific discovery. "I would like to take this opportunity to note my appreciation for the many lessons and wonderful gifts from yeast," Ohsumi noted. Showing his sense of humor at the Nobel Banquet, he added, "perhaps my favorite of all being sake and liquor."[173]

Autophagy has been highly conserved throughout the evolution of all cells that have subcellular organelles like nuclei and mitochondria (i.e., eukaryotic cells), from single-celled amoeba (some eukaryotes are unicellular) and fungi (e.g., yeast) to insects, reptiles, and mammals. It is an essential process that counteracts cellular injury and plays a critical role in maintaining homeostasis.

Once a tumor forms and begins to grow, cancer cells follow their own narcissistic aims, altering the signaling processes that control autophagy and upregulating the genes that drive this process. Amongst the genes regulating autophagy, the tumor suppressor p53 plays a critical role. By upregulating autophagy, tumor cells can cannibalize the components of the cell that are not required to drive growth and divert all the cell's energy and macromolecular resources to cellular proliferation. In response to chemotherapy, autophagy may enhance the survival of small numbers of tumor cells that manage to tolerate the damage and stress induced by cancer treatment, thereby providing these cancer cells with a survival advantage after exposure to toxic chemotherapeutic agents.

There is currently significant interest in learning how to turn down the autophagy rate in cancer cells. This is a rather challenging task given the "double-edged sword" of autophagy, which is both a tumor-suppressing and, under the wrong circumstances, a tumor-promoting biological process.[174] Although the task is daunting, it may not be beyond the capabilities of twenty-first century biologists armed with a map of the human genome and the tremendous computing power needed to take a serious run at unraveling and then genetically manipulating the enormously complex biological circuitry that regulates life at the level of cells and macromolecules.

From this information, a deeper understanding of life's biochemical processes continues to emerge, providing new avenues for both the prevention and the treatment of grievous human illnesses. As a result, new therapeutic tools continue to enter the fight against a vast array of grievous human illnesses, including heart disease, cancer, chronic inflammatory diseases, metabolic diseases (e.g., diabetes), and, last but certainly not least, debilitating infectious ailments—most notably, the newest entry in the pantheon of fearsome human diseases, COVID-19.

From the vantage point of a biochemist who entered the arena in the last quarter of the twentieth century, the medical potential of these discoveries provides exhilarating prospects for what might be possible in this century. As the nineteenth century was the era of chemistry and the twentieth will long be associated with astounding breakthroughs in physics, this new century will usher into human understanding and routine medical practice unprecedented biological wonders.

Chapter 8

The Ultimate Sugar High

The eukaryotic cell's metabolism encompasses an intricate panorama of biochemical reactions that convert food's molecular constituents and chemical energy into the building blocks of our macromolecules and the biochemical energy required to operate the cell.

The chemical building blocks of our macromolecules are obtained in two ways: (1) from the breakdown of ingested nutrients, and (2) by scavenging damaged macromolecules, harvesting their molecular constituents, and using these constituents as precursors for the biosynthesis of our proteins, lipids, carbohydrates, and nucleic acids.

Glucose and oxygen are the primary sources of biochemical energy in air-breathing life like us. The energy in the atomic bonds of foodstuffs is converted into biological energy in a process called *oxidative phosphorylation* (also called *aerobic respiration*). This energy is stored in a molecule called *adenosine triphosphate*, known by the acronym *ATP*. Most of the ATP in the cell is created during oxidative phosphorylation in the mitochondria, the cellular organelles often referred to as the "powerhouses of the cell."

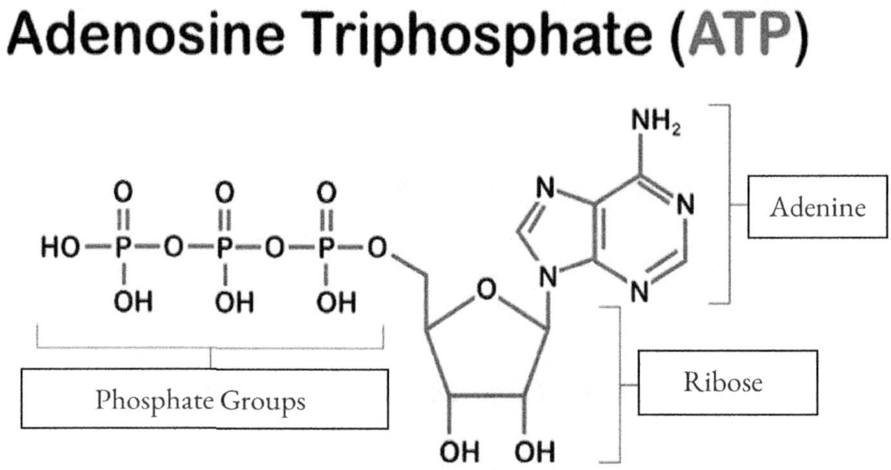

Adenosine Triphosphate (ATP)

Figure 12. The structure of ATP

    The ATP molecule can be divided into three parts (Figure 12). Adenine is one of the four nucleotide bases found in DNA and RNA; ribose is the sugar found in RNA (deoxyribose is the sugar in DNA); and, attached to ribose, three phosphate groups are strung together. The bonds between the phosphorus (P) atoms and the two oxygen (O) atoms that link the phosphate groups are high-energy bonds (called *phosphodiester bonds*).

    These bonds, formed by the action of enzymes, provide an efficient storage system for biochemical energy. When the bond is broken, energy is released to drive biochemical reactions in the cell. To release energy from ATP, the terminal phosphate group (the one farthest from adenine in the figure on the far left of the figure) is chemically removed (by reaction with water), thereby creating adenosine diphosphate (ADP). ADP is, in turn, phosphorylated by the enzyme ATP synthase, which adds a phosphate group to create a new ATP molecule to store more biochemical energy.[175] The energy that drives and sustains us is thus produced by simple biochemical reactions.

    This chemical mechanism has been so successful throughout

evolutionary history that it is ubiquitous throughout all forms of aerobic life. The 1997 Chemistry Nobel Prize recipient Paul Boyer elucidated this point in his acceptance speech for his work on the enzyme ATP synthase. Boyer noted that ATP's importance in the natural world cannot be overstated, as "the formation and use of ATP is the principal net chemical reaction occurring in the entire world."[176]

The cellular organelle responsible for most of the ATP formation, the mitochondrion (mitochondria is the plural), was first observed microscopically in the 1890s by Richard Altmann and Carl Breda. The linkage between cellular respiration and the creation of biochemical energy, in which the chemical energy in food is stored in the phosphate-phosphate bond in ATP in the presence of oxygen, was first made by German physician and chemist Otto Warburg in 1913.[177]

The glucose and oxygen levels throughout the body are carefully regulated to ensure homeostatic balance. Too little, the cells become quiescent, unable to muster the energy required for active metabolism. Too much can lead to cellular toxicities. For example, excessive levels of oxygen can lead to the overproduction of reactive oxygen species, which can damage our macromolecular constituents.

At the cellular level, glucose consumption is regulated by transmembrane proteins called *glucose transporters*. Once glucose enters the cell via its transporter, it engages in a series of enzymatic reactions, collectively called *glycolysis*, that break down the sugar molecule to harvest the energy in its chemical bonds. In addition, the carbon atoms contained in glucose are harvested to provide a carbon source for the biosynthesis of nucleic acids, amino acids, and lipids.

The glycolytic pathway involves ten enzyme-catalyzed reactions that form two, three-carbon molecules, called pyruvate, from each six-carbon glucose molecule. There are branch points in the glycolysis pathway that can divert the incoming energy and the carbon atoms in glucose to the formation of biosynthetic precursors for the cell's macromolecules. The diversion of glucose energy and carbon to biosynthesis occurs during times of stress, such as nutrient deprivation. Under homeostatic conditions, however, even in quiescent, non-proliferating cells, most of the pyruvate produced from glycolysis is used in the mitochondria in the oxidative phosphorylation process that generates most of the ATP in eukaryotic cells.

During oxidative phosphorylation, the energy-rich electrons present in the three-carbon molecule pyruvate are donated into a series of enzymatically driven biochemical reactions called the *Tricarboxylic Acid Cycle*, also called the *Krebs Cycle* after its discoverer, Sir Hans Adolf Krebs, a German-born British biochemist who worked out the cycle's steps in the late 1930s at Sheffield University. The Krebs Cycle uses electron transfer reactions, in which an electron donor is oxidized upon transfer of an electron to an electron acceptor (which is, in turn, reduced by the addition of the electron). These *oxidation-reduction reactions—redox,* for short—are used to harvest the energy in chemical bonds to generate cellular energy in the form of ATP.

The proteins involved in the electron transfer chain are found in the membranes of the mitochondria, where oxygen—so essential to aerobic life that its absence precludes the sustained generation of biological energy—is the final electron acceptor in the series of reactions leading to ATP formation. The electron transfer system is anchored in the mitochondrial membrane in an organizational motif like that found in the chloroplast, the organelle in plants that utilizes electron transfer reactions to generate glucose from carbon dioxide and water using sunlight (rather than the chemical energy stored in food) as an energy source.

The mitochondrion is an organelle with a fascinating backstory. It is widely accepted that the mitochondrion began life as a free-living microbe that merged with the eukaryotic cell early in the cell's evolutionary history. This mutually beneficial, symbiotic affiliation is called *endosymbiosis*, in which one organism exists as part of the other, for their mutual benefit, as opposed to traditional symbiosis, in which both participants are independent entities.

Over evolutionary time, the mitochondria retained their own DNA— called *mitochondrial DNA*, or *mtDNA*, for short. Mitochondrial DNA is separately replicated with respect to the rest of the cellular DNA. The fact that mtDNA has remained integral to the eukaryotic organelle suggests that this DNA, derived from the primordial mitochondria of long ago, is essential for proper mitochondrial function.

There are no mitochondria in sperm, which contain sufficient energy reserves for their mission of entering the vaginal canal in search of an unfertilized egg. The embryo's mtDNA is therefore entirely derived from the mother, passed along in the mitochondria present in the egg. You would, in fact, have a scientific basis to support the claim that you are a little more like your mom than your dad, genetically speaking.[178]

Homeostatic energy metabolism balances the need for the biochemical energy used to carry out the cell's functions with the need to always maintain a pool of molecular building blocks for repairing damaged macromolecules and, when necessary, for cellular growth (increase in size) and/or proliferation (cell division). The management of cellular energy stands in a delicate equilibrium: any energy used to manufacture macromolecular precursors depletes the cell's energy stores, precluding growth and/or replication; any energy used for ATP generation reduces the ability to build the stores of biochemical building blocks for macromolecular biosynthesis.

The atoms that comprise our macromolecules are derived from our food. These organic molecules—lipids, proteins, carbohydrates, and nucleic acids—contain only six types of atoms: carbon, hydrogen, nitrogen, oxygen, phosphorus, and sulfur.[179] These atoms are known by the acronym *CHNOPS*, pronounced like "schnaps," the German alcoholic beverage.

Two of the most important raw materials for producing our macromolecules are glucose and the amino acid glutamine. Glucose, as noted above, provides both biological energy and carbon for our macromolecules. Glutamine is a source of both carbon and nitrogen for macromolecular biosynthesis. Due to its criticality, glutamine is the most abundant amino acid in the blood and one of the seven essential amino acids that we must consume to survive. The other thirteen amino acids (the non-essential amino acids) can be biosynthetically manufactured by the cellular machinery.[180]

Like glucose, glutamine enters the cell via a transporter protein in the cell membrane. It is readily converted from glutamine to the amino acid glutamic acid (aka glutamate). Glutamate is a key neurotransmitter and the source of alpha-ketoglutarate, a related compound that participates in the Krebs cycle in the mitochondria.

Under homeostatic conditions, the cell's metabolic needs are balanced with the need to use its energy resources efficiently. Not so for the cancer cell, which is in the business of building more cancer cells. Except for quiescent cancer stem cells, most cancer cells rapidly metabolize glucose and glutamine to provide the energy and biochemical raw materials needed to continue expanding their numbers in their ruthless battle for survival and proliferation.

Otto Warburg was born in 1883 in Freiberg, Germany. As the son of the renowned physicist Emil Warburg—a friend of the most esteemed German physicist of his era, Albert Einstein—Otto grew up in comfortable surroundings. He was afforded the opportunity of pursuing the study of chemistry with famed chemist Emil Fischer in Berlin. In 1906, he earned a doctorate in chemistry, followed by a doctorate in medicine from Heidelberg in 1911.[181]

Warburg's research interest in the role of respiration in biological systems led him to study energy generation in tumors, focusing on the roles of oxygen and glucose consumption in cancer cells. Working first with slices of tumor tissue in the laboratory in the 1920s, Warburg made a profound discovery: he found that cancer cells rapidly metabolize glucose and produce significant levels of lactic acid. This indicated that tumor cells behave differently than normal cells in how they metabolize glucose.

In a subsequent series of cleverly designed experiments in living rats with laboratory-induced cancers, Warburg and his colleagues at Berlin's Kaiser Wilhelm Institute for Biology carefully extracted blood samples from the rats to follow glucose and lactate levels. By taking a sample from an artery leading into a tumor and collecting a sample from a vein exiting the tumor, it was possible to determine the tumor's impact on the levels of metabolites in the circulation. In control experiments (no tumor), the researchers demonstrated that the glucose and lactic acid levels were similar in the major artery and vein in several regions of the body.

In the experiments where the artery fed an established abdominal tumor in the rat, the data showed a significant decrease in glucose content, along with an increase in lactic acid, in the vein exiting the tumor.[182] Warburg noted, "The tumor takes, on average, 70 mg of glucose from 100 cc of blood, and returns 46 mg of lactic acid to the same amount of blood." This suggests, after appropriate calculations reflecting the molecular weights of glucose and lactic acid, about two-thirds of the glucose consumed was being converted to lactic acid, with about one-third of the glucose used for energy generation in the mitochondria.[183]

This was an important finding. In normal cells under homeostatic conditions, most of the glucose is used to fuel ATP production via oxidative phosphorylation in the mitochondria, a process that does not create lactic acid. These results showed that the metabolism of cancer cells with respect to glucose utilization was fundamentally different than that found in noncancerous tissues.

Warburg went on to postulate that cancer results from the conversion of glucose to lactic acid. This reaction is a type of fermentation, in which the pyruvate generated from the breakdown of glucose is converted to lactic acid to create energy rather than using the pyruvate to fuel the Krebs cycle during aerobic respiration in the mitochondria. Warburg concluded, "The prime cause of cancer is the replacement of the respiration of oxygen in normal body cells by a fermentation of sugar," such that "oxygen gas, the donor of energy in plants and animals, is dethroned in the cancer cells and replaced by an energy yielding reaction of the lowest living forms, namely, a fermentation of glucose."[184]

Warburg ascribed the predominance of glucose fermentation—the primary source of energy creation in prokaryotic organisms like bacteria—over mitochondrial respiration to a defect in the cancer cells' mitochondria. He hypothesized that the mitochondria of cancer cells are damaged, forcing them to ferment glucose rather than utilize this vital nutrient in the mitochondria as in normal cells.[185]

However, subsequent experimentation demonstrated that the mitochondria in cancer cells can create energy in the mitochondria, even though cancer cells metabolize most of their glucose through fermentation, notwithstanding the presence of abundant oxygen. This phenomenon—called *aerobic glycolysis* to specify that glucose fermentation occurs in the presence of oxygen—is a general feature of cancer cells. Aerobic glycolysis in cancer cells was subsequently named the *Warburg Effect.*

The Warburg Effect presents a scientific quandary: Why do cancer cells rely upon aerobic glycolysis? Lactate production is a highly inefficient means of energy creation compared to mitochondrial respiration. For each glucose molecule broken down for energy, oxidative phosphorylation generates about eighteen times more ATP than the conversion of pyruvate to lactate.[186]

The scientific consensus has (until recently) been that the Warburg Effect is an essential feature of cancer cells because it provides rapid energy and a selective survival advantage, as the inefficient fermentation process is faster than oxidative phosphorylation in the mitochondria. Recently, new data has suggested a far more intriguing rationale for the dominance of glucose fermentation in cancer cells.[187]

In addition to the need for energy, rapidly proliferating cancer cells also have a high biosynthetic demand to create the macromolecules required for building new cells. To fuel this need, cancer cells utilize a significant fraction of the incoming glucose to create biosynthetic intermediates for macromolecules. This condition favors fermentation as a means of rapidly breaking down glucose to generate energy, while also producing electrons during fermentation that support the redox reactions driving the biosynthesis of the macromolecules required to build new cells.

As a tumor grows and its need for nutrients and oxygen increases, the cancer cells are found in a state of nutrient depletion and oxygen deprivation (low oxygen is called *hypoxia*) that leads to additional metabolic adaptations by the cancer cells to continue their proliferation. As in many areas of their biological behavior, tumor cells demonstrate a plasticity of response in the regulation of cellular metabolism.

In the cancer cell, genetic and epigenetic regulatory mechanisms work together to "tune" the response to meet the needs of its local environment (called the *tumor microenvironment,* abbreviated *TME*). The TME contains multiple types of cells which, in turn, directly impact the fate of local cancer cells due to interactions (specified below) between the cancer cells and the other cell types in the TME.

It has been shown that "tumor cells utilize opportunistic modes of nutrient acquisition, which allow them to survive and proliferate in metabolically unfavorable conditions."[188] These adaptations, often stimulated by the action of oncogenic signaling and/or the loss of tumor suppressors, involve the activation of pathways leading to the acquisition of nutrients from sources that are generally unavailable in normal cells. They also involve the activation of survival pathways that provide for the ability to recover macromolecules from alternative sources when biosynthesis of the macromolecules is insufficient to meet metabolic demands.

While normal cells do not have a means for taking up macromolecules in the extracellular spaces, some tumor cells can engulf proteins through *macropinocytosis*, a process like phagocytosis of foreign bodies by cells of the innate immune system (Chapter 1). Under dire conditions, cancer cells are capable of *entosis*, the engulfment of entire living cells that are broken down by enzymes inside lysosomes, the membrane-surrounded cytoplasmic compartments laden with degradative enzymes. Cancer cells will even consume parts of themselves in a process called *self-autophagy*, in which macromolecules are liberated from cellular organelles and recycled into macromolecular building blocks. This process, a type of self-cannibalization, demonstrates the extremes to which cancer cells will go to continue their destructive path of uncontrolled proliferation.

For many patients, the metabolic dysregulation and smoldering inflammation characteristic of metastatic cancer lead to a severe and sometimes fatal consequence: a muscle-wasting disease called *cachexia*. Cachexia is powered by inflammation driven by the presence of elevated levels of pro-inflammatory cytokines such as TNF-alpha, IL-6, and IL-8 that are secreted by both immune cells and cancer cells.[189] The pro-inflammatory environment leads to biochemical and immunological defects that include the breakdown of fats and skeletal muscle.

The word cachexia, which translates from the Greek words *kakos* ("bad" or "worthless") and *hexis* ("state" or "disposition") as "bad condition," describes a state in which the breakdown of fats and proteins is accompanied by defects in the biosynthesis of these critical macromolecules.[190] The progressive loss of weight, skeletal muscle, and muscle strength leads to exhaustion. The syndrome is not fully understood, and therefore difficult to treat.

The wasting process is not reversible by consuming food or other standard nutritional methods used to treat starvation. Cachexia is a common feature of metastatic cancer; it is estimated that up to 80% of metastatic patients suffer from cancer cachexia, and about 25% die of cachexia due to respiratory or cardiac failure. This complex, multi-factorial disease is found not only in cancer patients, but also in people suffering from other inflammatory disorders, including chronic obstructive pulmonary disease (COPD) and rheumatoid arthritis (RA).

For aerobic creatures like us, oxygen is life. Without it, we cannot create the energy our cells need to operate the thousands upon thousands of biochemical reactions that sustain us. While oxygen's role in respiration, the process that produces most of our cellular energy, is widely appreciated, it may not be evident that oxygen also plays a critical part "in basic biological processes like DNA replication, transcription, translation, protein modifications, cellular signaling, and cell death."[191] Whether facing conditions of abundance or deprivation, oxygen plays a crucial role in the maintenance of homeostasis.

Aberrant oxygen signaling plays a significant role in a broad range of human diseases, "including cancer, cardiac disease, and neurological disease," maladies which all "have an underlying component of oxygen-based modifications of cellular and molecular machineries."[192]

Oxygen deprivation, like nutrient starvation, is a cause of cellular stress that is met with a physiological response designed to increase the chances of survival. In the absence of oxygen, cellular metabolism undergoes a series of rapid changes, including the shift from the oxygen-consuming process of mitochondrial respiration to glycolytic fermentation, which can occur in the presence of oxygen but does not require it for energy generation.

Transcriptional programs driven by the lack of oxygen are rapidly activated. As a result, the genes for protein messengers that serve as oxygen sensors known as *hypoxia-inducible factors—HIFs*, for short—are expressed to produce the HIF proteins: HIF-1, HIF-2, and HIF-3.

The active form of the HIF protein is a structure known as a *dimer*, in which two discreet polypeptide chains come together to form the active protein. A dimer has either two identical polypeptides or a pair of different protein molecules. In the case of HIFs, there are two different polypeptides, called *alpha* and *beta*, in the HIF dimer.

There are three types of alpha chains and one type of beta chain. Thus, three types of HIF protein dimers are possible, as each of the three types of alpha chains can pair with a single type of beta chain to form the active HIF protein.

Each polypeptide chain of the protein—each *subunit*, in the language of biochemistry—has a distinct role. The alpha chain is responsible for both sensing oxygen and acting as a transcription factor. The beta chain is a nuclear transport protein that, when stably paired with the alpha chain, can transport the alpha chain into the nucleus. There, the HIF protein binds to promoters called *hypoxia-responsive elements* on the DNA to induce the expression of hypoxia-sensitive genes.

The alpha chain's oxygen-sensing mechanism is mediated by a biochemical modification that regulates the biological response mediated by the HIF alpha chain. In the presence of sufficient oxygen, enzymes called prolyl ("PRO-LEEL") hydroxylases add hydroxyl groups (-OH) to the amino acid proline in the alpha chain. This action facilitates the destruction of the alpha chain through a carefully regulated biochemical process with a tongue twister of a name, *ubiquitination* (pronounced "u-bik-wit-in-ā-shun"), in which an enzyme attaches a protein called ubiquitin to a protein targeted for destruction. Addition of ubiquitin to the protein serves as a biochemical flag signaling that the target protein is destined for degradation so its components can be recycled for the biosynthesis of new proteins.[193]

Following ubiquitination, the targeted protein undergoes further biochemical events that result in its destruction in a cytoplasmic structure called the *proteasome*. The proteasome, a large protein complex, is responsible for breaking down the peptide bonds in proteins (a process called *proteolysis*) so that the constituent amino acids can be harvested for new rounds of protein synthesis.

When oxygen levels drop, the biological activity of the proline hydroxylases falls, such that the enzymes serve as highly adept oxygen sensors. "A 10% decrease in the oxygen concentration has demonstrated a rapid lowering in the hydroxylation of proline residues on HIF-alpha," thereby stabilizing HIF-alpha so that it can associate with HIF-beta and subsequently enter the nucleus to exert its transcription-factor activity."[194]

During hypoxia, the binding of the HIF proteins to their promoter sites on the DNA generates waves of genetic expression that allow the cell to cope with hypoxic conditions. This survival mechanism ensures the conservation of energy and molecular resources in the absence of the ATP generated by mitochondrial respiration at physiologically normal oxygen levels.

HIF expression supports the shift from oxidative phosphorylation to the fermentation of glucose resulting from the Warburg Effect. The changes elicited by HIF include the induced expression of variants of the enzymes responsible for glycolysis, favoring those variant forms that support the Warburg Effect. In addition, HIF expression induces an increase in the expression of glucose transporters to boost the entry of glucose into the cell. This provides a vast increase in the rate of glycolysis to generate metabolic intermediates supporting the continued proliferation of tumor cells.

As a result of the increase in glucose fermentation to lactic acid, the TME becomes acidic, a condition known as *acidosis*. Whereas the pH of the blood that feeds our cells ranges under homeostatic conditions from 7.2-7.4 (slightly basic), the elevated levels of lactic acid produced inside cancer cells induce upregulation of membrane transport proteins that extrude protons from the cell to raise the intracellular pH to physiological levels (the higher the proton concentration, the lower the pH).[195] This process decreases the pH of the extracellular environment surrounding the tumor (a pH of 7 is neutral; below this is called acidic, and above, basic). Measurement of the extracellular pH using sophisticated imaging techniques has shown that the pH in the TME can go as low as 6.44.[196] The acidity in the TME leads to alterations in the metabolic profile of cancer cells that help drive cancer progression.

The metabolic reprogramming induced under hypoxic conditions provides the cell with the best chance of survival in the face of acidosis and the lack of oxygen. HIF-induced expression in the normal cell is a temporary stress response to adverse conditions. This response has evolved for short-term, emergency applications only, so the cell can either recover or, in short order, perish. It has not evolved to go on indefinitely.

The environment created by cancer cells is one of biochemical chaos, a five-alarm fire from the perspective of the integrity of the cell's homeostatic biological functions. Since molecular oxygen is not readily consumed by mitochondrial respiration due to the Warburg Effect, the life sustaining O_2 is extensively converted to ROS, thereby sustaining the hypoxic state in the TME. Under these hypoxic conditions, the levels of reactive oxygen species (such as superoxide) skyrocket.

During glycolysis, the enzymatic reactions responsible for harvesting carbon from glucose for the biosynthesis of macromolecules also generate ROS. The Warburg Effect in cancer cells promotes rapid rates of glycolysis and fermentation, adding to the ROS overload and further sustaining hypoxia. Finally, since the activity of the prolyl hydroxylases is inhibited by hypoxia, the depletion of O_2 by ROS formation prevents HIF-1 alpha degradation. As a result, the HIF-1 alpha-beta dimer continues to activate HIF-1 gene expression.

The HIF protein plays another key role in the cellular response to hypoxic stress. The HIF-alpha subunit, when bound to its beta subunit partner, induces the expression of genes that enrich and stabilize stem cell populations. This process ensures the presence of the appropriate stem cell populations to replace cells damaged by hypoxia. Because cancer stem cells are present in most cancers, the HIF pathway plays a critical role in promoting tumor progression. As cancer cells evolve "stem cell-like" characteristics, they acquire the genetic plasticity needed to propel them along the road to tissue invasion and metastatic spread.

It has long been appreciated that rapidly dividing cancer cells consume considerable amounts of nutrients. Most significantly, they ravenously consume glucose to fuel cell division and to provide the metabolic intermediates for the biosynthesis of the macromolecular components of the growing population of cancer cells. Since the delivery of nutrients to cancer cells is reliant on sufficient blood flow into the tumor, cancer cells manipulate signaling pathways to increase their blood supply. The biological process responsible for blood vessel formation is called *angiogenesis*. Specifically, angiogenesis refers to the formation of new capillaries derived from existing ones in a process in which a small "budding" protrusion on the outer surface of a capillary wall develops into a new capillary.

In a 1945 paper, Glenn Algire and Harold Chalkley compared the capillary growth elicited by tumor cells implanted under the skin of cats with that resulting from the implantation of normal cells. They found that the implanted tumor led to a far more significant level of new blood vessel formation than when normal cells were implanted in the same location. The data suggested "that the growth of a tumor is closely connected to the development of an intrinsic vascular network."[197]

By the 1960s, a body of evidence had accumulated showing that not only are tumor cells able to elicit the formation of new blood vessels, they also are able to induce this effect at a far greater distance than implanted normal cells (both of which require new vasculature for survival).

The biochemical mechanisms underlying the angiogenesis process became the lifelong fascination of a Harvard-trained surgeon named Moses Judah Folkman. The son of a Boston rabbi who had seen his father ministering to the spiritual needs of disease sufferers, Folkman was a young surgeon-in-training in the early 1960's at the National Naval Medical Center in Bethesda, MD, attending to more worldly needs in his work for the Navy. His goal was to find a workable solution to the battlefield problem of rapid blood loss.

Sterile solutions containing the oxygen-carrying protein hemoglobin were under development as a blood substitute that might stabilize critically wounded sailors on battleships and aircraft carriers until they could be airlifted to a hospital for intensive care treatment. The experimental system involved freshly prepared canine thyroid glands, through which the hemoglobin solution was passed (*perfused*, in biological terms) to determine the ability of the hemoglobin solution to stabilize the perfused gland. The hemoglobin perfusion was effective, such that the glands were viable for up to two weeks. In addition to the hemoglobin solution's ability to maintain cell survival, its ability to enable cell growth was also shown.

In these experiments, mouse melanoma cells were injected into perfused glands and their growth was measured. Sure enough, perfusion supported the ability of the thyroid glands to support the growth of the melanoma cells. During the melanoma cell implantation experiments, Folkman observed that when the tumors grew to about 1-2 mm in diameter, they stopped growing. He also noted that no new blood vessels had formed near the small, growth-arrested tumors.

Looking to determine the biological characteristics of these non-vascularized, tiny tumors, Folkman and his team carefully extracted the small tumors from the perfused glands and implanted them into the host mice from which the melanoma was derived. Following implantation, the small tumors grew rapidly and became vascularized as new branches budded off nearby capillaries. By inducing the formation of new blood vessels in the host animal, a tumor that had been arrested at a tiny size could be reenergized to become a proliferative growth. Folkman had arrived at a fascinating conclusion: the development of new vasculature is required for tumor growth. Stated more simply, tumor growth beyond a tiny size requires a new blood supply.

Thinking that tumor vascularization must be a biochemically regulated process and that a biological signal must be required to induce blood vessel formation near the tumor, Folkman postulated "that tumors express angiogenesis-inducing molecules required to orchestrate the angiogenesis process."[198] The requirement for an angiogenesis "trigger" would explain why tumor growth is limited, and angiogenesis absent, in a perfused gland. The canine thyroid would not respond to angiogenesis-inducing factors released by mouse tumor cells. However, when the mouse tumor was re-implanted in the mouse, the angiogenesis-signaling molecules would be recognized, and the living, quiescent tumor arrested in the canine gland would come back to life.

The obvious corollary struck Folkman immediately. If tumor growth is angiogenesis-dependent, it might be possible to arrest cancer growth by blocking angiogenesis in the vicinity of the tumor, thereby preventing the proliferation of tumor cells. As Folkman noted in a 1971 theoretical paper in the *New England Journal of Medicine*, "tumor growth is angiogenesis-dependent and the inhibition of angiogenesis could be therapeutic."[199]

This idea formed the nexus of a brilliant scientific career in which Judah Folkman pioneered a global effort to understand the role of angiogenesis in cancer. That effort continues today. Over a 40-year career, Folkman's contributions to the field included the isolation, from a rat breast tumor cell line, of the first biological molecule with angiogenesis-promoting properties—which, in turn, resulted in a pioneering contribution to the field of anti-angiogenesis tumor therapy.

As a result of Folkman's work, a dozen angiogenesis inhibitors have been developed to date for clinical use. In addition, his research efforts provided the biological knowledge required for developing Avastin, an antibody that blocks an angiogenesis-inducing protein called *Vascular Endothelial Growth Factor* (*VEGF*). Avastin, the trade name for an antibody called bevacizumab, was licensed in 2004 by the biotechnology company

Genentech (a subsidiary of Roche) for the treatment of colorectal cancer and is now licensed for multiple cancer indications.

At the time of Folkman's prescient insights, the widespread belief was that tumor cells grew along pre-existing blood vessels. So widely believed was the existing scientific dogma about the source of the tumor's blood supply that Folkman's first grant proposal to the National Cancer Institute (NCI) on the study of tumor vascularization was rejected on the grounds that "It is common knowledge that the hypervascularity associated with tumors is due to dilation of blood vessels and not new vessels and that this dilation is probably caused by the side effects of dying tumor cells."[200] The closing argument of the NCI rejection demonstrated a level of certainty that was evidently far in excess than warranted. It claimed, "tumor growth cannot be dependent upon blood vessel growth any more than infection is dependent upon pus."[201]

The analogy misfired, as the NCI reviewers had, in this case, fallen into the trap of putting too much faith in the current dogma. This tendency can result in the inability to objectively evaluate a new proposal that happens to contradict what you think is widely known.[202]

Folkman's story is illustrative of how innovative ideas often meet heavy resistance. This has been a common theme in the history of scientific discovery, illustrated by the intense skepticism and scrutiny he faced from the scientific community until the importance of tumor angiogenesis became widely accepted in the 1990s. Today, the inhibition of angiogenesis still plays a significant role in the treatment of many cancers.

Chapter 9

The Wound That Does Not Heal

Born in 1821 in the Prussian farming village of Schivelbein (currently the town of Świdwin in northwestern Poland), Rudolf Carl Virchow lived in an age when Prussian medicine had not yet reached the same level of sophistication as that found in England and France. In those countries, medical scientists recognized that microscopic analysis of living organisms provides an essential foundation for medical understanding. German scientists Matthias Schleiden and Theodor Schwann first proposed the cell as the basic building block of life in the late 1830s. This result was the product of Schleiden's studies of plant cells in 1838, followed by Schwann's studies of animal cells in 1839.[203] These discoveries were contemporaneous with Virchow's medical studies, which began at the Prussian Military Academy in 1839. Consequently, the cell theory (which, ironically, has German origins) had no real impact on Prussian medicine at the time of Virchow's medical training.

After graduating in 1843, Virchow discovered scientific literature from other European countries. He keenly appreciated the idea that the key to understanding human biology was to understand what goes wrong when cells are in a diseased state. The fact that cells are the building blocks of life was known at the time, but how the cells interacted with each other was a mystery. Virchow encouraged his students to "learn to see microscopically," to take full advantage of microscopic instruments to probe the secrets of biology at the level of the cell.[204]

By marrying macroscopic observations of patients with microscopic analysis, Virchow forever altered how diseases are diagnosed and treated. His most important contributions to medicine include the discoveries that leukemia is an abnormality of white blood cells, that inflammation is at the root of atherosclerosis (hardening of the arteries), and that emboli (e.g., blood clots and fatty deposits) can cause pulmonary disease.[205] As a result, Virchow redefined our understanding of cell biology: beyond being mere building blocks, each cell was the descendent of another cell, and these cells worked together to maintain health. Human disease was not due to sickness of the entire person; there were no humors capable of causing bodily illness, as many believed. Rather, our ailments, he believed, are due to changes in specific cells in the body.

Virchow's legacy includes the science of pathophysiology, the study of the changes in cellular function at the root of disease. The ramifications of this shift in medicine are evident when we see a pathology report from a biopsy or other surgical procedure.

The Prussian physician's interest in diseased cells led him to the investigation of cancer. He noticed that human tumors contain leukocytes. The presence of white blood cells suggested that these cells, known to be present in healing wounds and absent from healthy tissues, play a role in cancer.[206]

In the middle of the nineteenth century, physicians knew that cancer sometimes develops at the site of a wound where proper healing has not occurred (for example, in the stomachs of people who suffer from chronic ulcers). Virchow also noted that the characteristic overgrowth and tissue damage (scarring) seen in wound healing is reminiscent of similar processes observed in cancer progression. Based on the similarities between damaged tissues and biopsied tumor samples, Virchow proposed, in 1863, "that chronic irritation and previous injuries are a precondition for tumorigenesis."

A hundred years later, British scientist Sir Alexander Haddow proposed that "tumor production is a possible over-healing."[207] American biologist

Harold Dvorak further refined this idea into a pithier phrase by suggesting in a 1986 *New England Journal of Medicine* paper, "tumors are wounds that do not heal." He proposed that "in contrast to healing wounds, the process is not self-limiting in cancer tissue, resulting in uncontrolled cell proliferation, invasion, and metastasis."[208]

When a tissue undergoes an injury, the immediate reaction to the insult is the activation of the coagulation cascade, a series of biochemical reactions mediated by proteins that ensure rapid clotting in wounded tissues to prevent blood loss. The presence of the clot elicits a response from immune cells that subsequently perform the wound-repair process.

The first types of immune cells that arrive at the wound are neutrophils, the front-line soldiers of the innate immune system. These fascinating and multi-functional cells are ever vigilant to the presence of imminent threats to homeostasis. Neutrophils are the most abundant type of leukocyte in the blood. The trillions of neutrophils found in human blood comprise up to 70% of the white blood cell count. It is estimated that the bone marrow produces 100 billion neutrophils per day.[209] Circulating throughout the tissues in search of microbial and viral invaders, neutrophils also traffic to other regions of inflammation, such as damaged tissues, where they assist in cleaning up cellular debris prior to the initiation of the repair process.

These cells are well-equipped for their assigned functions. They are not only phagocytic, capable of engulfing microbes and their cellular debris; neutrophils are also potent generators of reactive oxygen species that can damage the biochemical integrity of infiltrating organisms. In addition, they have a unique and fascinating property: they can undergo a process called *netosis*.

Netosis is a form of apoptosis in which a neutrophil discharges a complex of DNA, histones, and antimicrobial proteins from the cell that can entrap and kill microorganisms and viruses. The DNA-protein net projected outside the neutrophil during its final act of self-destruction is called a *neutrophil extracellular trap* (*NET*). Dysfunctions in NET signaling have been implicated in cardiovascular diseases, autoimmune diseases, and cancer progression.[210]

Expanding on Dvorak's insight, the actions of immune cells in wound healing are highly reminiscent of what takes place in the presence of a solid tumor. From inflammation's continuous role throughout tumor development and progression to the specific activities of the immune cells that participate in these processes, cancer is a disease driven by the same inflammatory mechanisms observed during the repair of injured tissue.

Thus, cancer may indeed be a wound that does not heal. It may, in fact, behave like a wound that continues trying to heal while rendering the opposite result. Instead of working with other cells in a coordinated fashion (as in wound healing), cancer cells free themselves from the regulatory controls active under homeostatic conditions. As a result, the wound repair systems operate in a perverse manner that increases cellular damage rather than repairing it.

From ROS-induced genetic damage to the activation of proto-oncogenes and loss of tumor suppressors, to the corruption of biological signaling processes, to the participation in (and encouragement of) a state of molecular chaos, cancer cells continue along their evolutionary path toward metastatic spread. Along the way, they co-opt non-transformed cells, such as those of the immune system and even the normal cellular constituents near the cancer mass, to facilitate the tumor's progress along its destructive path.

A mountain of evidence demonstrates the importance of inflammatory processes in cancer. Current estimates indicate that about one in five human cancer deaths can be attributed to the effects of chronic inflammation. Inflammatory sources include chronic bacterial and viral infections, chronic obesity (which causes the release of pro-inflammatory cytokines), chronic exposure to environmental pollutants such as pesticides, alcohol, tobacco smoke, heavy metals (e.g., chromium and lead), and radioactive contamination. All can contribute to the formation of dangerous reactive oxygen species, as well as *reactive nitrogen species* (*RNS*)—which, like ROS, are highly reactive compounds that can damage our macromolecules.[211]

The most recent data indicates that cells of the immune system are present in the tumor mass over the entire lifespan of cancerous tissues, from tumor initiation through the development of metastatic disease. This continuous inflammatory process is now recognized as a general feature of cancer.[212] Inflammation's role in cancer is so complex that it is still a major focus of cancer research. Even with the advent of new and promising immunotherapeutics for metastatic cancer, many questions remain about the role of host immunity in cancer.

Before the emergence of modern genetics, it was difficult to explain how gametes perform the natural miracle of creating new life. Without information about the organization of the genetic material in the form of the DNA double helix, it was impossible to formulate a mechanism that explains how the combination of the egg and sperm gives rise to living progeny.

Near the end of the nineteenth century, German biologist August Weismann proposed that the fertilized egg is the source of all the diverse types of cells that comprise the fully developed organism. He also proposed that some of the genetic information contained in the fertilized egg must somehow be removed during the development of the specialized cells that comprise the body's tissues. This meant that the specialized, differentiated cells are fundamentally altered from those in the early embryo, such that the somatic cells lack the genetic components contained in the gametes, which are required for the formation of the entire organism. In this manner, a mechanism (later called the *Weismann Barrier*) prevented somatic cells from functioning as gametes: "As cells progress along their various pathways of differentiation, genes no longer required for other divergent lineages are cast off or permanently inactivated."[213]

In the twentieth century, microscope technology advanced rapidly, and it became possible to perform cellular microsurgery to remove the cell's nucleus and place it into an enucleated egg—an egg with its nucleus removed. This technique, called *somatic cell nuclear transfer* (*SCNT*), can place the nucleus from an early embryonic cell into an enucleated egg to determine whether the embryonic cell can provide all the genetic information required to grow a new organism.

In 1952, Robert Briggs and Thomas King, at Philadelphia's Fox Chase Cancer Center, successfully obtained live tadpoles following the transfer of the nucleus from an early embryonic cell of the frog *Rana pipiens* into an enucleated frog egg.[214] The outcome of the experiment demonstrated that the cells of the early embryo contain the entire complement of genetic information required to develop a new organism.

The next step was to see what happens when a differentiated somatic cell is transplanted into an enucleated egg. It was conceivable that a fully differentiated, tissue-specific and highly specialized cell lacking the plasticity observed in stem cells might be altered by the biochemical processes that take place in the embryo during development. After several years working the kinks out of the challenging experimental manipulations involved in SCNT, British biologist John Gurdon (Oxford University) transplanted the nucleus of a somatic frog cell into an enucleated egg of the African horned toad *Xenopus laevis* and produced viable offspring.[215]

These results were astonishing. The viable birth of the fully developed organism from the nucleus of a somatic cell indicated that due to its presence in the biochemical milieu of the egg, the somatic cell was fully reprogrammed to act like a germ cell (aka a gamete). Somehow, the chromosomal information contained in germ cells that programs the development of the organism, although inactive in somatic cells, is available for expression if somatic cells are placed in the proper milieu.[216]

This landmark achievement showed that although cellular differentiation is a unidirectional path from stem cells to specialized cells under homeostatic conditions, it is not inherently a one-way trip. In addition, the information in the gametes is not erased from or permanently disabled in the genome of somatic cells. This information may be offline in somatic cells, but it is still present. Amazingly, the nuclear material from the somatic cell can be reconfigured to operate like the DNA present in the gametes. Gurdon's work verified that August Weismann's hypothesis about gene inactivation during the differentiation of somatic cells was correct, with the exception that the genes are not permanently lost in somatic cells.

By 1982, mice were successfully produced from a somatic cell and enucleated mouse egg using SCNT. In 1996, at the University of Edinburgh, a team led by Professor Sir Ian Wilmut successfully produced Dolly—presumably the most famous sheep in human history—using the SCNT method.[217] In the current century, scientific advances have built upon the foundation laid by SCNT by removing the egg from the process and employing mixtures of transcription factors to reprogram cellular differentiation in the test tube to create pluripotent stem cells from somatic cells.

In these cells—known scientifically as *induced pluripotent stem cells*, or *iPSCs* (the "i" indicates that the cells are derived by biochemical manipulation)—the differentiation process has been reversed, from specialized cells to pluripotent stem cells. Thus, in the presence of the right concoction of gene-activating proteins, a specialized cell can be sent back "up" the differentiation pathway all the way to powerful pluripotent stem cells capable of generating, through sets of successive stem cells, all the specialized cells in the organism.

The discovery that somatic cells can be reprogrammed into cells with the phenotype of stem cells demonstrates that the cell has innate plasticity, the ability to alter its transcriptional patterns under the influence of sets of master transcription factors that can direct the state of cellular differentiation. The 2012 Nobel Prize was awarded to two cell biologists, British scientist Sir John Gurdon and Japanese biologist Shinya Yamanaka, for their discoveries showing how somatic cells can be reprogrammed into induced pluripotent stem cells using four transcription factors. By treating tissue-specific differentiated cells using this method, it is possible to create stem cells that can generate every cell type in the body.

The medical applications of a technology offering this capability are beyond comprehension and far beyond the scope of this narrative. In brief, we can envision treating diabetes by implanting insulin-secreting islet cells in diabetic patients (this work is already well underway) or implanting liver cells that manufacture factor VIII (clotting factor) in hemophiliacs. It will be a fascinating—and, no doubt, controversial—scientific (and societal) story to watch over the coming decades.

We have known for over half a century that histone modifications play a role in gene expression. The correlation of acetylated histones with regions of genomic activity and the accompanying biochemical evidence supports the hypothesis that the modification of histones has functional significance in gene expression. Such modifications alter the three-dimensional structure of the chromatin itself by causing small perturbations in the overall chromatin architecture.

These conjectures from long ago turned out to be correct in principle. However, half a century ago, we missed the mark on appreciating the extent of chromatin complexity. We did not yet understand the fundamental importance of chromatin's three-dimensional architecture on gene function. In addition, our thoughts were focused on structural changes along small stretches of the chromatin fiber; that is, we were mostly thinking about local structural effects.[218]

We knew about transcription factors in the 1970s but didn't realize that some of these proteins, the primary TFs, can impact gene expression over extensive regions of the chromatin. We certainly had no knowledge of long-range regulating elements, such as enhancers.

Five decades hence, we have a far greater understanding of the complexity of the biochemical modifications that impact chromatin structure and function. We also have come to appreciate that chromatin's three-dimensional structure is every bit as important to its function as the three-dimensional structure of a protein is to its activity.

Considering the emerging picture of chromatin architecture, and the role of epigenetic modifications in its regulation, we return to the DNA airport. First, we imagine a three-dimensional object, like a rope, oscillating in a sine wave pattern, with equal distance between successive peaks and successive valleys, a distance we will call (analogous to electromagnetic radiation such as light) the wavelength. Now imagine that while one region of the wave pattern remains fixed, the wave pattern compresses in an adjacent region (smaller distance between crests and troughs of the sine wave) to a wavelength one-tenth the size of the original distance. We will call the original pattern the "open" configuration and the compressed pattern the "closed" configuration.

Our ropes, the runways in our DNA airport, are laden with activity and, in addition, the structure is twisting and gyrating in three-dimensional space. This alone would make aircraft operations quite difficult. Still, we need to add the consideration that there are differences in the levels of tarmac congestion between the open and closed regions. While both have activity, the closed regions have additional barriers, mediated by structural modifications, that further restrict access to and from the tarmac.

Turning back to cancer cells, data shows that they can directly modify the chromatin architecture to open regions that are in the closed state in a normal cell of the same lineage. In this manner, the cancer cell can activate genetic activity that operates at other points of the life cycle (for example, during embryogenesis and/or wound repair), leading to the expression of genetic programs at the wrong place and wrong time, thereby promoting metastasis instead of ensuring homeostasis.

In 2016, Sarah K. Denny and her colleagues at Stanford reported, in small cell lung cancer cells in culture, a transcription factor called *NFib* that can "reconfigure" chromatin and render previously inaccessible areas into a more open configuration, thereby driving the metastatic properties of the cells.[219] Moreover, when the genomic architecture of the DNA taken from the isolated metastatic lesions of lung cancer patients was compared to that of their tumor counterparts in the primary lesion, the results showed that "about 24% of the accessible regions in the metastases" had a higher degree of open chromatin architecture. The altered regions were found not in protein-coding sequences but rather in regulatory elements that control gene expression in these open chromatin regions.[220]

The ability to alter genetic expression patterns in response to environmental cues is vital for cellular survival. The cell must be able to react to stimuli by choosing from the menu of responses available in the active portions of the genome. While a cancer cell (or a stem cell) has more choices on the menu than a fully differentiated cell, all cells can respond to the challenges of environmental changes via the cellular plasticity that is lurking, to one extent or another, in the genome of every cell.

This plasticity provides for the generation of increasing numbers of cancer cells that differ from each other due to the mutations and adaptations the cells undergo in response to the tumor microenvironment. Therefore, as a small tumor grows, distinct subpopulations of cancer cells (genetically related cell clones) diverge from the original cancer cell as well as from each other.

Each clonal variety has inherent genetic and epigenetic variability because of mutations and other changes to the DNA, driving a relentless competition for oxygen, nutrients, and macromolecular precursors. As conditions in the tumor bed become increasingly non-homeostatic, the cancer cells are engulfed in a life-or-death competition. This battle takes the cancer cell on a long, perilous journey from birth to its inevitable demise, either because the cancer cell cannot compete with its rivals, or because it can compete and is fated to perish upon the death of its host.

According to the stem cell theory of cancer, different subpopulations of cancer cell clones with stem cell-like properties are in a fierce competition for survival. These cells demonstrate significant plasticity, with the ability to respond rapidly to environmental stresses to generate cells capable of surviving unstable conditions and fierce competition in the tumor microenvironment. This ability provides another supporting argument for the stem cell theory, as growing cancers and metastatic lesions are well known for adapting quickly to thwart therapeutic intervention.

Additional compelling evidence for the cancer stem cell (CSC) theory is derived from experiments showing that the transcriptional programs of the stem cell populations of tumors resemble those of the stem cells found in the tissue of origin. For example, "the CSC program of mammary carcinomas resembles in outline the corresponding program in the normal mammary gland."[221]

While differentiation in living organisms is considered, from a "classical" viewpoint, a one-way path from cells that demonstrate significant plasticity to those that are more "fixed" in their behaviors, recent evidence suggests that, at least in the context of cancer stem cells, there may be cells capable of undergoing "de-differentiation" similar to that seen in the laboratory using techniques such as SCNT.

Several laboratories, including that of eminent cancer researcher Robert Weinberg at MIT, have published data demonstrating that "not all cancers adhere to a unidirectional CSC model." The evidence demonstrates the presence of "subpopulations of non-CSCs that could readily switch from the non-CSC to CSC state (with concomitant epigenetic changes) in response to environmental stimuli and genetic manipulation—such as the presence of transcription factors and/or microRNAs that regulate pluripotency and are often over-expressed in human cancers."[222]

The capability of a cell to go "backward" on the path of differentiation, to undergo "de-differentiation," had not been confirmed previously. While this ability cannot be positively ruled out for normal stem cells, no evidence supports this. However, the ability of specific differentiated cells to morph into cells demonstrating stem cell behaviors in the context of cancer appears to be a certainty. As in the case of metabolism and energy utilization, the cancer cell can upend the "rules" of cell biology for the processes that regulate the determination of cellular identity. In so doing, they warp that sense of identity and distort the purpose of the cells from which they are derived.

These findings suggest the possibility that within living tumors, cancer cells are transforming from stem-like cells to non-stem-like cells and back again, perhaps many times throughout the lifespan of the tumor. Under homeostatic conditions stem cells can only be generated from other stem cells. The ability to generate stem cells from non-stem cells (if proven) would be a magical power in the eyes of classically trained cell biologists. Such a capability would allow a tumor to generate a virtually limitless supply of stem cells to continue feeding its growth.

The ramifications of bi-directional differentiation in tumors include a reassessment of the nature of tumor progression. In addition to the model where a cancer stem cell gives rise to a small number of additional cancer stem cells along with a large number of the differentiated cancer cells that comprise the bulk of the tumor mass, we need to add a new model. In this new model, the cancer stem cell population is in some form of equilibrium with the non-cancer stem cell population, with cells switching back and forth between stem-like and non-stem-like states.

There is an additional consideration in the model regarding the origin of cancer. In addition to the idea that a normal stem cell can give rise to a cancerous stem cell, it is also possible, in the context of bi-directional differentiation, that cancerous stem cells can arise from differentiated cells that have been transformed into a cancerous, more "stem-like" state.

These considerations have significant implications regarding our understanding of cancer and how we might combat its spread. We need to consider a greater number of pathways to tumor initiation and progression. With more pathways to consider, the range of therapeutic approaches also widens, offering innovative approaches to treatment as we broaden the potential targets for eliminating tumor-propagating and treatment-resistant tumor cells.

As we shall see in the following chapters, there are definitive signs of progress in the quest to leverage our growing knowledge of cancer biology in the management of this dreadful human affliction that touches so many lives.

Chapter 10

The Seed and the Soil

Death caused directly by the primary tumor is rare. It can occur when the growth of the primary mass impinges on vital structures. In about 90% of cases, cancer mortality results from the slow, progressive transition of a primary tumor to the deadly disease we call *metastatic cancer*.

While we tend to think of cancer in the long-held duality of the site where cancer first appears (the primary tumor) and sites where new tumor masses have arisen beyond the location of the primary tumor (*metastatic lesions*), this view is oversimplified when we take into consideration cancer's physiological complexity. As a tumor progresses, its constituent cells become increasingly enmeshed in the body's overall physiology, co-opting normal cells to support the tumor in its deadly business of growth and invasion. In this sense, metastatic cancer is a disease of the entire body system—a *systemic disease*, in the language of medicine—and its effects on the body are widespread and profound.

The connection between primary tumors and their metastatic cousins has been recognized since the early nineteenth century. The word metastasis was first coined—as *metastase*, in French—by surgeon J.C.A. Recamier (1774–1852).[223] Recamier, a highly accomplished surgeon who was drafted into the army of the Alps during the French Revolution, was the author of one of the earliest medical textbooks on oncology. Published in Paris in

1829, Recamier's treatise provided detailed accounts of 85 cases of breast cancer along with observations on cancers at 13 other sites in the body.[224]

By the mid-nineteenth century, cancer investigators (including Rudolf Virchow) hypothesized that metastasis occurs when pieces of the tumor break off the tumor mass, either as cells or clumps of cells called *emboli*. These cells then enter the circulation through either the blood vessels or the lymphatics.

The lymphatic ducts that carry cells of the immune system throughout the body drain into lymph nodes. Here, immune cells (or cancer emboli) can enter the bloodstream. At the time, scientists believed that once constricted and rendered immobile in tiny capillary beds, cancer emboli form metastases. Therefore, the sites of metastatic growth from a given tumor were a matter of random chance determined by the location of arrest of the tumor cells in capillaries.

In 1889, English surgeon Stephen Paget investigated whether any patterns could be discerned in the distribution of metastatic lesions in breast cancer patients. He studied the autopsy reports of 735 women who had perished from metastatic breast cancer and found a marked preference for metastasis to certain organs. Metastases were found mainly in the liver, ovary, and specific bones, with a far lower preference for the spleen.[225] This finding established that, at least in the case of breast cancer, the distribution of metastatic lesions could not be explained by random arrest of cancer cells in the circulation. As he noted in his 1889 paper, "remote organs cannot be altogether passive or indifferent" [226] when it came to the sites of arrest of tumor cell emboli. Reaching for a botanical analogy, Paget noted, "When a plant goes to seed, its seeds are carried in all directions, but they can only live and grow if they fall on congenial soil."[227] Paget proposed that the development of a metastasis depended not only on the tumor cells, but it also depended on the biochemical characteristics of the location in the body where the tumor cells landed. In botanical terms, the outcome depended not only on the seed, but also on the soil.

While the existing model for metastasis recognized the role of the seed, Paget now expanded the model to include the importance of the soil.

Paget's insight is known as the *"seed and soil" hypothesis*. In more modern parlance, the hypothesis proposes that *disseminated tumor cells (DTCs)* that leave the primary tumor and are deposited in a microenvironment favorable to their growth. This idea remains a fundamental principle of our twenty-first century understanding of metastasis.

Current data suggests that the successful implantation and growth of metastatic lesions is a low-probability event. Tumor cells undergo a perilous journey in the circulation in the face of host immunity. Estimates suggest that the survival rate of tumor cells that enter the bloodstream is less than 0.1% (1 in a thousand). Thus, only a tiny fraction of the tumor cells in the circulation survive long enough to arrest in a distant site where, if conditions are favorable, they can grow into viable tumors.[228] Even if a cancer cell survives in the circulation, the odds of successfully implanting in congenial "soil" are also vanishingly small. As a result, very few disseminated tumor cells, less than one in a million, have a chance of forming a growing metastasis.

How does a cancer cell detach itself from the primary tumor and begin its journey to distant sites in the body?

To understand how cancer cells leave the site of the primary tumor, we can start by considering how our tissues are organized, and how our cells are programmed to move throughout our body when required, such as during embryogenesis and wound repair.

All epithelial cells—the cells that line all the body's external and internal surfaces—grow in flat sheets, with each cell connected to its neighbors by surface proteins that form junctions with the surface proteins of the adjacent cells. The junctions provide structural stability for the epithelial tissue, along with the ability to rapidly change cell-to-cell contacts by modulating the protein-protein interactions at the intercellular junctions.

The epithelium's stability is maintained by the surrounding *stroma*, the connective tissue and blood vessels that surround the organs. The stroma fills the intercellular spaces with loosely packed cells intertwined with the extracellular matrix (ECM). Fibroblasts, the most abundant cell type in the ECM, are mini-protein factories that express substantial amounts (in cellular terms) of the extracellular matrix proteins that provide a three-dimensional protein framework for the organization of sheets of epithelial cells. In addition to its role in the maintenance of the structural architecture of our organs, the ECM proteins modulate cell movement in response to the secretion of signaling molecules by fibroblasts.

The stroma's cellular components are known as *mesenchymal cells*. Mesenchymal cells (like all cell types) are derived from embryonic stem cells. During fetal development, the ESCs give rise to mesenchymal cells with the ability to detach from the epithelial sheet (through repression of production of the proteins that form the intercellular junctions), as well as the ability to move from the epithelium through the ECM and beyond. To make this possible, the epithelial phenotype (appearance and behaviors) of subsets of the early embryonic cells is genetically programmed to change to cells with migratory characteristics.

The transition of the cells of the early embryo from the epithelial to the mesenchymal phenotype is mediated by specific biological signals that initiate the expression of a set of master transcription factors. These master TFs, which have widespread impacts on genetic expression via their interactions with enhancer sequences on the DNA, have cool names like SNAIL and ZEB. These DNA-binding proteins activate genes that confer mesenchymal characteristics upon subsets of epithelial cells, allowing them to detach from the cellular colony and migrate to new locations.

This programmed change in cellular phenotype is called the *epithelial-to-mesenchymal transition* (*EMT*). This process is active during both embryogenesis and wound healing. In the former, the EMT takes place to enable cellular migration to form new tissues and organs. During wound healing, the EMT is activated to allow for the migration of cells that fill the locations previously occupied by damaged cells.

To close the wound following its repair by the inflammatory response, epithelial cells adjacent to the wound undergo the EMT and migrate inward toward the wound field. Once they have occupied their new locations, these cells undergo a reverse transition, from the mesenchymal phenotype to that of an epithelial cell. This allows them to reintegrate into the epithelial sheet by establishing the intercellular junctions characteristic of epithelial cells. This process is called (logically) the *mesenchymal-to-epithelial transition* (*MET*).

The biological signaling that mediates this reversible phenotype involves not only the transcription factors themselves, but also sets of microRNA species that modulate these transitional events. By interacting with the TFs, the microRNAs impact the stability of the TFs. The microRNAs are derived from non-coding RNA transcripts that are spliced in diverse ways to provide several iterations of the miRNAs, each of which has a specific (and unique) impact on the stability of the transcription factors.

In this way, a small number of transcription factors and an array of microRNAs derived from a small number of transcripts provide a highly effective and efficient means for regulating the activity of transcription factors. Such a regulatory network can respond to changing conditions rapidly, regardless of the levels of transcription factors present, simply by changing the composition of the microRNA pools. This regulatory process is highly efficient with respect to the utilization of raw materials and energy.

Pretty neat.

Most cancers, around 80%, are derived from the epithelial cells in the parenchyma, the non-connective tissues that comprise the organs. Epithelial-derived cancers, known collectively as *carcinomas*, include the human cancers with the highest occurrence rates, such as most skin, breast, lung, colon, liver, and stomach cancers. The remaining 20% of cancers—the sarcomas—are derived from precursor cells of mesenchymal origin, such as cells found in the muscles, bones, fat, connective tissues (like cartilage), nerves, and circulatory system. Since carcinomas comprise the majority of human cancers, most of our knowledge on cancer cell behavior is derived from studies of epithelial cancers.

For metastasis to occur in epithelial cancers, some of the cancer cells must leave the primary tumor and move into the vasculature to spread through the body. A carcinoma cell in the primary tumor can acquire the ability to detach from the epithelium and migrate out of the ECM. To do so, it must acquire mesenchymal characteristics. The transition of a carcinoma cell in the primary tumor to a DTC involves the same EMT program that occurs during embryogenesis and wound healing.

The biological milieu inside a growing primary tumor is staggeringly complex. There is significant heterogeneity in gene expression and epigenetic patterns amongst the individual tumor cells in the population. The EMT and MET programs are not "on-off" switches that render one phenotype or another. Rather, these genetic programs elicit a continuum of biological characteristics for each cell. When we characterize the population of tumor cells that leave the tumor bed, we find both epithelial- and mesenchymal-related "markers" (proteins) present on their surfaces. This finding suggests "that an active EMT process is likely to operate during the dissemination of cancer cells from the primary tumor."[229]

While it appears that many of the DTCs migrate as individual cells, clusters of disseminating cells are also present in the circulation of some cancer patients. The involvement of mobile clusters of cells, a process called *collective migration*, has been documented in embryogenesis, wound repair, and, most recently, cancer.[230] While both individual tumor cells and tumor cell clusters are capable of serving as migratory seeds for metastasis, it is not clear which modality is more capable of seeding metastatic growth.[231]

Once the cancer cells/clusters have broken free of the primary tumor and invade the surrounding stroma, they continue their migratory path toward the circulation. To move through the stroma, migrating tumor cells utilize a mesenchymal characteristic required for moving through tissue: the ability to secrete protein-degrading (aka *protease*) enzymes (called *matrix metalloproteinases, MMPs*) that degrade other proteins, such as those of the ECM. MMPs are also active in the homeostatic process of wound healing during the remodeling of damaged tissues.[232]

The DTCs enter the blood and/or lymphatic vessels in a process called *intravasation*, in which signaling elicited by the tumor cells render the vessels more susceptible to DTC entry into the circulation. This signaling causes the endothelial cell lining of the blood vessels to become "leaky" and more permeable.

Once in the circulation, the DTCs interact with platelets, blood cells that are critical in the generation of clots in regions of damage to the vasculature. This interaction can shield the DTCs from recognition by immune cells (such as macrophages, natural killer cells, and T cells). In addition, platelet binding can assist the tumor cells in adhering to the walls of the vasculature, arresting them in the blood vessel at a location where they might leave the vasculature (in a process called *extravasation*) through the action of MMPs secreted by the tumor cells. Extravasation opens a path for the DTCs to penetrate the vessel wall and enter the tissues of an organ distant from the primary tumor bed.

Several factors play a role in determining the fate of disseminated tumor cells that enter new sites. These new sites, called *pre-metastatic niches*, are biochemically groomed for future implantation of DTCs by signaling mechanisms between the growing primary tumor and the normal cells in the niches. This signaling establishes fertile ground for the transplantation of tumor cells that have survived the perilous journey from the primary tumor mass. These signaling mechanisms involve both the cytokine network and signaling proteins called *chemokines*, which are attractant molecules that elicit a migratory response (sometimes at a considerable distance) in other cells.

Additionally, inter-cellular signaling between the tumor cells and normal cells in both the tumor microenvironment and distant pre-metastatic sites is orchestrated by "packets" of lipids, proteins, and RNAs called *exosomes* (*exo* means "outside" or "external"). A type of structure called an *extracellular vesicle*, exosomes are used by normal cells for intercellular transfer of molecular information. They are comprised of a spherical lipid membrane containing proteins, microRNAs, and/or mRNAs. On the exterior face of the lipid membrane, proteins that bind to cells with a specific surface receptor protein provide a targeting mechanism allowing this form of biological communication to act both locally and at distant sites throughout the body.[233]

Tumor cells actively utilize the transfer of molecular information via exosomes to drive tumor progression. Both tumor cells and stromal cells in the TME that have been co-opted by the tumor cells secrete exosomes containing a payload of biomolecules that enhance cancer progression. The molecular information provided by the exosomes prepares the pre-metastatic niches for successful implantation and growth of disseminated tumor cells. Once the pre-metastatic niche has been biochemically pre-conditioned, DTCs in the circulation are attracted to these sites by chemokines secreted by cells in the pre-metastatic site.

The information contained in the TME-derived exosomes helps drive tumor progression. As a result of exosome activity, tumor cells undergo an increase in *cell motility* (the ability to move in three-dimensional space). Biological signaling from both tumor and stromal cells in the TME provides additional fuel for tumor progression, including an increase in the growth of blood vessels to feed the tumor bed and an enhancement of immunosuppression by myeloid-derived suppressor cells.[234]

Once transplanted to a pre-metastatic niche—and thereby establishing what is called (obviously) a *metastatic niche*—the tumor cells of the new malignant lesion are faced with the biological challenges of an unfamiliar environment, with conditions that likely vary significantly from those of the primary tumor. The metastatic cells must adapt to their new home, where they face differences in the components and biological characteristics of the immune infiltrate, the presence of a unique cytokine network, a changed metabolic profile, and a different stroma, with its own unique composition and biological signaling profile.

If the newfound tumor colony is unable to establish a firm foothold in the new soil and fails to adequately adapt to establish a growing colony, the transplanted cells can enter a state of biological quiescence—a survival mode of existence—that is called *tumor dormancy*. Herein lies one of the secrets of cancer's biological durability, a key to its malignancy: the adaptability of cancer cells in response to cellular stress provides a survival capability not available to normal cells.

Here we also find one of the major reasons why achieving a long-term remission in metastatic cancer is so difficult. Like many viruses, dormant tumor cells can remain in a latent state for prolonged periods of time, probably many years, until the moment arrives when conditions become favorable to enter into a proliferative state.[235] Furthermore, because our current therapeutic arsenal against cancer—surgery, drugs (including most monoclonal antibodies), and radiation—all target growing cancer cells, dormant metastatic cancer cells are highly resistant to these forms of therapy. Tumors knocked down by therapy to a level where it is difficult to detect their presence can reappear years later—often with intensified malignancy compared to the original cancer.

When we characterize the dormant DTCs, we find that they resemble stem cell populations. They are both capable of long-term survival by entering a quiescent state, awaiting changed conditions until they become activated to yield progeny cells of distinct types. They also both demonstrate significant cellular plasticity, the ability to change their phenotypes when stimulated to do so. Finally, they can both establish new colonies of growing cells.

While we do not have the scientific capability to define dormant DTCs as actual stem cells, we can say they are "stem cell-like" or, alternatively, that they have "stemness." And it is this very stemness that provides them with their power to resist our therapies and stifle our attempts to eradicate them.

Cancer's astounding biological complexity presents a significant barrier to formulating a comprehensive and systematic description of cancer cells and their behaviors that can be applied across the many varieties of human cancer cells. At the end of the twentieth century, biochemists Douglas Hanahan (University of California, San Francisco) and Robert A. Weinberg

(MIT) published a landmark paper called "The Hallmarks of Cancer." The paper was premised on their belief they could "foresee cancer research developing into a logical science, where the complexities of the disease, described in the laboratory and clinic, will become understandable in terms of a small number of underlying principles."[236]

The "Hallmarks" paper provides a biological framework for understanding the properties of cancer cells under the rubric of six "functional capabilities" that describe the observed behaviors of cancer cells. The 2000 "Hallmarks" publication was followed up by the authors in 2011 by an article with a title containing a clever salutation to Star Trek (at least I see it that way): "Cancer Hallmarks: The Next Generation."[237] The latter publication added two "emerging hallmarks" and two "enabling characteristics" that further describe the behavior of cancer cells. Taken together, the hallmarks and enabling characteristics describe both the behaviors of the cancer cells themselves as well as the interactions of cancer cells with non-cancerous cells during tumor progression.

The ten functional capabilities of cancer cells described by the hallmarks and enabling characteristics (Figure 13) cover a broad range of biological functions. The capabilities describe the behaviors—the phenotypes that can be observed—of cancer cells in living tumors. The observed behaviors driven by these capabilities can also be characterized in the realm of cells and molecules by describing the underlying biochemical mechanisms that drive the phenotypes of cancer cells. These observed phenotypes can be grouped into three underlying biochemical mechanisms that drive the hallmark behaviors.

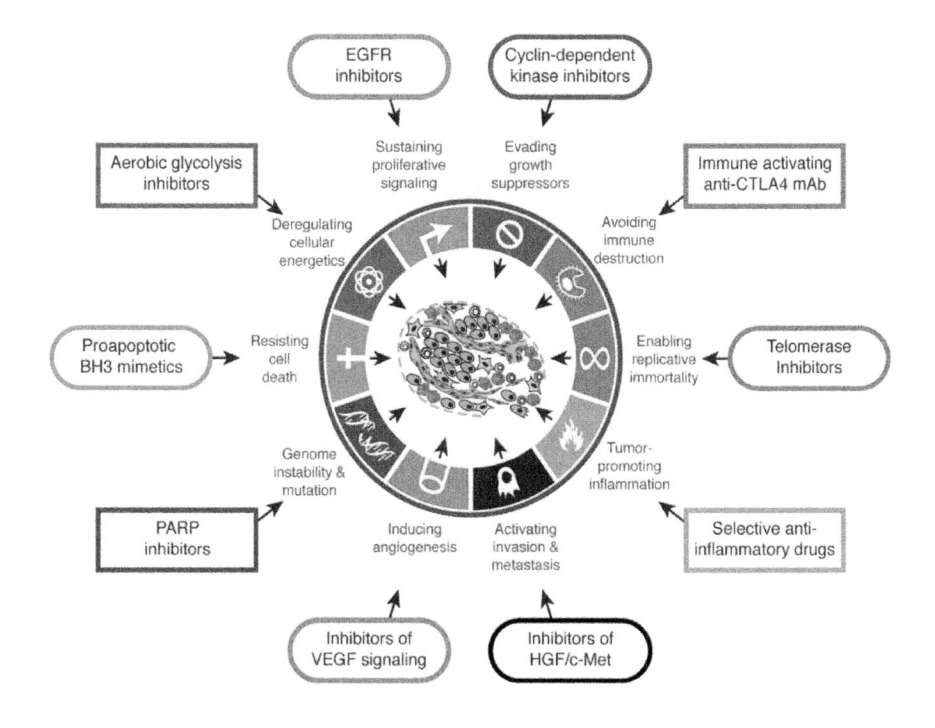

Figure 13. The Hallmarks of Cancer

From: D Hanahan and RA Weinberg, *Hallmarks of Cancer: The Next Generation*. Cell 144(5):646-74 (2011). The figure also shows potential therapeutic interventions that address the hallmarks and enabling characteristics.

These biochemical mechanisms work together to co-opt existing cellular systems and utilize them to support uncontrolled proliferation and tumor progression. The three mechanisms—which will be referred to here as the *"mechanistic pillars of cancer"*—involve alterations in three key cellular processes, whereby cancer cells and the tumor cell network, working with the (unwitting) cooperation of normal cells, enable key biochemical processes that drive cancer progression. The three proposed mechanistic pillars are:

Mechanism 1: Disruption and co-option of cellular signaling networks to favor cancer cell proliferation, enable the avoidance of immunosurveillance, induce angiogenesis, and resist apoptotic and other cell death signals. The ability of cancer cells to act independently while also engendering the cooperation of non-transformed cells has its origin in the subversion of existing signaling pathways to those that favor cancer cell survival and tumor progression. Cancer cells are secret agents—hardened by their arduous battle for survival and growth—that corrupt cellular communications and manipulate cell signaling for their own nefarious purposes.

Mechanism 2: Reprogramming of the patterns of genetic expression in a manner that enhances genetic instability and promotes cellular plasticity, ensures replicative immortality, and supports invasion and metastasis. Cancer cells can leverage, to their own advantage, the innate plasticity of the genetic mechanisms of the cell to alter their phenotypes. By reprogramming the genome via changes in the expression patterns of transcription factors and microRNAs, alterations in the epigenetic landscape, and modulation of chromatin architecture, cancer cells can (in the vernacular of our times) hack the programs responsible for executing genetic expression to favor their own survival and proliferation.

The corruption of gene expression in cancer cells includes their ability to access sections of the biological programming that are off limits under homeostatic conditions. This is how cancer cells can render upon themselves the ability to perform cellular functions normally reserved for stem cells during embryogenesis and wound healing. The stem cell-like properties of cancer cells support the migratory behavior that leads to invasiveness and eventual metastasis, the capacity for self-renewal, and the cellular plasticity that provides an extraordinary level of adaptability to changing conditions.

In this manner, cancer cells are out of step with the organism's life cycle. There is nothing inherently malignant about the programs of genetic expression rendered by the cancer cell. They are simply activated in the wrong place, at the wrong time, and for the wrong purposes from the perspective of the survival of the cancer patient.

Mechanism 3: Exploitation of biochemical pathways to promote non-homeostatic conditions that drive chronic inflammation and changes in cellular metabolism. The resulting aberrant environment of low oxygen (hypoxia) and elevated levels of lactic acid (acidosis) favors rapid proliferation of cancer cells under unstable conditions that accelerate the evolution of cancer cell clones and favor the competitive success of cancer clones of increasing metastatic potential.

By leveraging existing biochemical pathways—such as those used during embryogenesis and wound healing—cancer cells both promote and take advantage of non-homeostatic conditions, thereby creating an environment of molecular chaos that is highly aberrant with respect to homeostatic conditions. This environment is one of metabolic dysregulation and chronic inflammation, redolent with a toxic brew of reactive oxygen and nitrogen species, hypoxia, and acidosis. Under these conditions, the cancer clones compete for survival while promoting a state of rapid cellular evolution in the highly mutagenic environment of the cancer niche.

These three mechanisms underlie the phenotypes that develop during cancer progression. Each of the capabilities described in the "Hallmarks" model is driven by one or more of the mechanisms. The illumination of the molecular wizardry in the cancer cell's repertoire provides insight into why achieving long-term remissions in patients with metastatic cancer is so difficult. On the other side, we can see that by understanding these characteristics and how they relate to cancer progression, and by investigating the mechanisms that support cancer phenotypes, we can uncover new therapeutic avenues that exploit these characteristics for the benefit of cancer patients.

We will explore these innovative approaches in the following chapters.

Chapter 11

Hard to Kill

The long-used, traditional modes of cancer therapy—surgery, chemotherapy, and various forms of ionizing radiation—are highly effective at reducing the size of tumors, often achieving significant reductions in tumor mass. In early, localized tumors, there is a possibility that surgical excision of the tumor, properly executed to ensure the total removal of the lesion with appropriate cancer-free margins under microscopic examination, may be sufficient to obtain a long-term remission.

In invasive cancer, where tumor cells have exited the primary tumor, obtaining a long-term remission is more difficult. Given the accelerated rate of clonal evolution in cancer cell populations, resistant clones inevitably develop. Even when the therapeutic intervention is achieving significant reductions in or even the apparent elimination of tumor cells, there is always the possibility that a dormant clone that survived the therapeutic onslaught might one day return to a proliferative state.

Having survived therapy, these quiescent cells remain under the radar, invisible and undetectable. Since they are inactive, they lack the metabolic signature of cancer cells and therefore cannot be detected by existing diagnostic techniques, which depend on the ability to monitor the metabolic signature of cancer (such as rapid glucose uptake). When a once-dormant clone awakens to refuel tumor progression, the survivor clone is generally more virulent and destructive than its predecessors.

Even when performed by the most experienced practitioners, surgical excision of a primary tumor, without evidence of tumor invasion or spread, cannot guarantee a long-term remission. It should be noted that evidence supports the contention that surgical intervention may, in and of itself, facilitate the migration of cancer cells away from the tumor bed. The probable cause of cancer cell dispersion from the tumor during surgery can be found in the nature of surgery itself, which disrupts the integrity of the tissues and vasculature in the region of the tumor. Surgical disruption may expedite the passage of tumor cells into adjacent tissues or even into the bloodstream or lymphatic system.

For this reason, surgery is often followed by the mainstays of cancer therapy: toxic drugs and/or ionizing radiation. Chemo- and radiotherapy are directed at any residual cells in and near the tumor bed that escaped excision during the surgical procedure along with those that migrated from the tumor bed into nearby lymph nodes. While it has long been believed that cells capable of invasion and metastasis do not appear early in the life cycle of tumors, recent evidence suggests the probability that invasive cells may begin their long and perilous journey from the tumor bed at an early stage in some cancers.

The traditional treatment modalities, chemotherapy and radiation, are designed to interfere with specific cellular processes. For example, many chemotherapeutic agents target DNA replication, DNA repair, or other processes critical to control of the cell cycle. Radiotherapy is based on the principles first described over a century ago by H.J. Muller, who demonstrated in fruit flies that X-rays cause mutations due to DNA damage. When cells undergo extensive DNA damage inflicted by toxic agents, the genomic damage impairs DNA replication, resulting in cell death by apoptotic signaling. The presence of the dying cells can, in turn, stimulate the immune system to recognize and destroy tumor cells that survived the initial treatment.

The use of ionizing radiation in cancer treatment is based on the presumption that cancer cells are deficient in their DNA repair and therefore cannot recover from the genetic damage inflicted upon them during treatment. In contrast, normal, untransformed cells that are fully functional in their DNA repair can recover from this damage. This therapeutic principle also applies to chemotherapeutic agents that target DNA replication and repair.

The presence of intact caretaker and gatekeeper functions in normal cells ensures that cells with DNA damage that cannot be repaired get a "failing grade" from the cellular quality control system and are culled from the herd prior to replication. Such quality control functions are often deficient in cancer cells due to the activation of oncogenes and the loss of tumor suppressors. Because of inadequate quality control, cancer cells with damaged DNA can proceed through the cell cycle and pass their damaged DNA on to daughter cancer cells. This capability, however, is finite. DNA damage can become so extensive that the cell passes its "breaking point," the point at which the damage is too severe for cell survival.[238]

Since chemotherapy and radiotherapy target critical cellular proliferation processes, they may not significantly impact cancer cells in a non-proliferative, quiescent state. Unlike proliferating cells that are taking up nutrients to build metabolic intermediates for the biosynthesis of macromolecules for new cells and replicating their DNA, dormant cells have a low metabolic rate. They do not incorporate appreciable amounts of cellular toxins and are therefore far more likely than proliferating cells to survive therapeutic interventions targeted at disrupting cancer cell proliferation.

The duration of the quiescent state that cancer stem cells can undergo can be measured in years and perhaps even decades. Herein we find an explanation for the extended periods sometimes observed between an apparent cure and cancer's return decades later.[239] These cancer stem cells are "time bombs" that are significantly more stem-like, invasive, and aggressive than their predecessors. Here we also find another reason reversing metastatic cancer is amongst the most challenging clinical problems in medicine.

For decades, the therapeutic focus of the cancer treatment strategy has been on poisoning the cancer cells to a greater extent than normal cells and hoping the normal cells can mount a comeback from the toxic insult. In the process, normal cells are also directly impacted, and their functionality can be diminished. This includes the cells of the immune system, which are essential for orchestrating the response by the host defenses. In addition, toxic therapies can contribute to genetic instability by causing mutations and other forms of DNA damage. This damage, in turn, contributes to the genetic instability that helps drive tumor evolution.

The second weakness of this therapeutic strategy is that cancer cell populations readily develop resistance to treatment. Therapeutic resistance is responsible for the low rate of long-term effectiveness of many cancer treatments. Even in the case of the targeted biological therapies that emerged in the 1990s, such as monoclonal antibodies, the development of therapeutic resistance is common. As the cancer cell population continues along its evolutionary path, mutations inevitably produce clones that can overcome the impediment rendered by the targeted agent. In cases where early intervention renders significant signs of success—including the apparent eradication of the primary tumor, and even in cases where metastatic lesions have been diminished— therapeutic resistance can develop over time.

While there is yet much to learn about why this is so, our knowledge of the biochemical properties of cancer cells provides insights into the development of therapeutic resistance. We can start with the heterogeneity of the cancer cell population. This population is under myriad selective pressures due to continuously changing conditions, such that cancer progression is an evolutionary process driven by Darwinian selection.

Due to chronic inflammation, hypoxia, acidosis, and the presence of significant levels of reactive oxygen species (ROS) in the tumor microenvironment (along with genetic instability), the TME is constantly undergoing biochemical changes. Throughout the tumor bed (and in pre-metastatic and metastatic niches, if present), the levels of ROS, oxygen, nutrients, and metabolites vary from place to place. Thus, the tumor microenvironment itself is a collection of micro-niches that provide a tremendous diversity of conditions for the evolution of cancer cell clones during tumor progression.[240]

The highly heterogeneous cancer cell population and variable microenvironments in the tumor bed present a staggeringly complicated medical challenge in the effort to achieve long-term remissions. Even when a therapeutic agent is effective against some clones, it likely will not similarly impact the entire population of cancer cells given the likelihood some cancer clones will evolve the ability to resist the treatment.[241] The evolution of resistant clones is inevitable: "Because of large-scale genomic alterations and consequent diversity, the emergence of resistance is predictable as a fundamental property of carcinogenesis itself."[242]

It has taken a half-century of scientific inquiry and medical experience to learn that targeting tumors by aiming to destroy rapidly growing cancer cells is an inadequate long-term treatment strategy. Repeatedly, we have seen that if we wish to eliminate the cancer cells, we need to go to their source. We have learned that cancer cells, like their non-transformed counterparts, are part of a cellular lineage containing stem cells (or, at the very least, "stem-like" cells) that are quiescent, with low rates of proliferation and cellular metabolism.

The inevitable conclusion is that the future of successful cancer therapy rests on innovative approaches that target all the cancer cells, including the stem cells, or, even better, selective agents that specifically target and destroy the stem cells. There might be medical advantages to the latter approach, including the avoidance of a dangerous complication of cytotoxic therapies called *tumor lysis syndrome* that can severely impact multiple body systems and, if left untreated, cause organ failure and death.

Tumor lysis syndrome is a dangerous side effect that results from the release of the intracellular contents of the large population of dead and dying cells (both cancerous and normal cells alike) present following chemotherapy and/or radiation treatment. The bolus of intracellular contents released into the bloodstream causes metabolic changes that damage the kidneys and other organs. Tumor lysis syndrome sometimes occurs in cancer patients spontaneously in the absence of cytotoxic therapy.

Cancer biologists classify therapeutic resistance mechanisms in two broad categories: intrinsic and extrinsic resistance. The former includes changes that occur to the cancer cells themselves. For example, mutations that alter the structure of the therapeutic target (a protein) may prevent the drug from binding to cancer cells. In addition, genomic changes that alter cellular communication networks allow cancer cells to "bypass" the pathway(s) impacted by the drug. Cancer cells can also reduce the numbers and/or binding properties of the surface receptors targeted by cancer drugs, thereby impacting the efficacy of cancer therapeutics.[243]

Extrinsic resistance refers to processes that alter the tumor microenvironment to support tumor cell resistance to the therapeutic agent. For example, in the presence of the metabolic chaos and smoldering inflammation wrought by the tumor, normal cells in the tumor stroma secrete proteins that promote tumor progression. Such proteins include cytokines, proteases, and growth factors.[244]

These bystander cells are responding appropriately to the presence of an anomaly in their microenvironment—namely, the presence of a wound in the form of a tumor. They respond by activating inflammatory processes designed to protect the body from harm. These processes break down damaged cells (e.g., autophagy) and thereby generate the macromolecules needed for tissue remodeling during wound repair.

As a result, the tumor cells are provided with what they need to continue their incessant march toward metastasis. This includes growth factors that support cellular proliferation and other protein factors (such as transcription factors) that inhibit tumor cell apoptosis, promote an inflammatory tumor microenvironment, sustain the growth of new vasculature, and contribute to changes in the microenvironment that promote tumor cell invasion of the tumor stroma. Such changes include the secretion (transport of proteins through the cell membrane into the extracellular space) of degradative protease enzymes and signaling molecules that disrupt the junctions between epithelial cells. In this manner, the non-tumor cells in the TME are victimized by the tumor cells in their midst, transformed into unwitting participants in their own destruction.

Some cancers are dependent on the continued expression of oncogenes for cancer progression. In such cases, we say that the cancer is "oncogene addicted." A type of breast cancer called HER2 positive (HER2+) breast cancer provides an example of an oncogene-addicted tumor. HER2 is a member of the same family of growth factor receptors as epidermal growth factor receptor (EGFR). Epidermal growth factor (EGF) stimulates the proliferation of the epithelial cells lining both the external (skin) and internal surfaces of our organs. HER2, the key growth factor involved in the development and maintenance of breast tissue, is over-expressed in about 20% of breast cancer patients.

When the HER2 receptor is present in appreciable amounts, breast cancer patients can be treated with an antibody to the HER2 receptor that blocks its activity. This monoclonal antibody, called trastuzumab and marketed under the trade name Herceptin, is widely used to treat HER2$^+$ breast cancer. Monoclonal antibody drugs like Herceptin that block cellular growth factors offer, in some cases, impressive and durable results.

These innovative cancer medicines are subject to some of the same barriers to their therapeutic efficacy as conventional chemotherapy drugs, including the development of resistance due to the biological trickery characteristic of evolving cancer cells. Because redundancy in the biological circuitry provides multiple ways to obtain responses in many (if not most) biochemical circuits, cancer clones inevitably develop that can bypass the biological pathway targeted by the drug.

In addition, there are several ways for the cancer cell to overcome the inhibition on survival and proliferation exerted by a therapeutic agent. The cancer cell may evolve the ability to reduce the uptake of the drug (described below), and/or the cancer cell can leverage existing emergency response genes to ensure cell survival and replication. In some cases, a cancer cell can express enzymes capable of detoxifying the chemical agent.

Given the heightened genomic instability characteristic of cancer cells, mutations can also alter the efficacy of therapeutic agents. For example, as noted previously, a mutation in the molecule targeted by a drug might reduce or eliminate the recognition of the target by the therapeutic agent.

This increases in probability with time as the genetic instability of cancer cells increases during cancer progression.

Due to the dynamic nature of the genomic changes in cancer cells and the profound acceleration in cellular evolution characteristic of growing tumors, cancer treatment is an arduous battle against an unpredictable and elusive target.[245]

A molecule (including a toxic agent) can enter a cell in three ways. Most simply, the molecule can pass through the cell membrane, a process known as *diffusion*. The cell membrane is comprised of two overlapping layers of lipids (a lipid bilayer) studded with membrane proteins. To diffuse through the membrane, a molecule must have chemical properties that enable it to interact with the membrane's components in a manner that permits free passage of the chemical into the cell.

Due to the low probability that the chemical interactions between substances and the membrane's components will allow free passage into the cell, most chemicals do not diffuse through the membrane. Exceptions include water, carbon dioxide, and oxygen, all of which can freely diffuse through membranes.[246] The passage of ions and other charged substances across the membrane (e.g., sodium, potassium, chloride, magnesium, calcium, etc.) requires the activity of *membrane transport proteins* (aka *transporters*), which shepherd chemical species found in living organisms across their membranes.

Transporters are proteins that span the cellular membrane and form a channel for the entry and/or exit of specific molecules into or out of the cell. An abundant family of transporters in eukaryotes is called the *ABC transporters*. ABC stands for ATP-binding cassette. These proteins move molecules across the membrane, both into (influx) and out of (efflux) the cell, using ATP as an energy source. The most common ABC transporter is a *glycoprotein* (a protein molecule decorated with sugars at specific amino acids) called *P-glycoprotein*, one of about fifty known ABC transporters in humans.[247]

Like the other ABC transporters, P-glycoprotein interacts with an array of compounds that are either neutral or positively charged. The ABC transporters are not selective for a single molecule. Biochemists call a protein with this characteristic a *promiscuous protein* (and not due to its sexual proclivities). Due to this promiscuity, P-glycoprotein can interact with many of the commonly used anti-tumor agents.[248]

Many cancer cells express elevated levels of P-glycoprotein, and that the transporter expressed in cancer cells can bind to chemotherapeutic agents and prevent them from entering the cell.[249] Because of the promiscuous nature of P-glycoprotein binding, the transporter's elevated level of activity in cancer cells can provide resistance to multiple drugs. The expression of P-glycoprotein (and other members of the ABC transporter family) can therefore result in multi-drug resistance; the gene that codes for P-glycoprotein is called *MDR1* (with *MDR* standing for *multiple drug resistance*).

Another mechanism whereby molecules in the environment can be taken up by cells is a process called *endocytosis*. In this process, a section of the membrane encapsulates material outside the cell and incorporates the material into the cell's cytoplasm in the form of a *vacuole*, a membrane-surrounded compartment in the cytoplasm. The general term endocytosis incorporates several distinct processes. The cell can incorporate a molecule bound at the surface (*receptor-mediated endocytosis*), consume an entire organelle or even a cell (phagocytosis), or "drink" the fluid outside the cell to incorporate molecules in the extracellular space (*pinocytosis*).[250]

While capable of killing cancer cells, chemotherapy and radiation also damage healthy tissues. The more aggressive the chemotherapeutic regimen, the more damaging it is on normal cells. However, these treatments can prevent recurrence of disease for years following surgery, and sometimes even provide long-term (even lifetime) remissions. Thus, these mainstays of cancer therapy meet the quintessential definition of a "double-edged sword."

Regardless of the patient's outcome, chemotherapy and radiation treatment cause massive cellular damage, creating a vigorous inflammatory response. In addition to activating the inflammatory response, damage to the macromolecular components of the tumor cells' microenvironment activates damage responses that are a natural reaction to cellular injury in the tissues. These "damage and stress-response programs have evolved to prevent the propagation of oncogenic genetic damage to progeny."[251]

For cancer cells, the response to cellular stress that protects normal cells provides fuel for the tumor's fire, accelerating cancer cell evolution and establishing an environment favorable to metastasis.[252] The pro-inflammatory environment stimulates angiogenesis and favors cancer cell survival by reducing apoptosis and supporting cellular proliferation. As a result, the tumor microenvironment is conditioned by the natural survival responses of the bystander stromal cells to create an environment that promotes cancer cell progression and provides sanctuary for the cancer cells from subsequent treatment.[253]

The DNA damage inflicted on both the cancer cells and the normal cells by genotoxic therapies such as traditional chemotherapy and radiation activates the DNA Damage Response. While normal cells can repair most of the DNA damage, the deficiencies in DNA repair in cancer cells make them far more tolerant of DNA damage at the checkpoints designed to protect the genome's integrity.

Following treatment with genotoxic agents (chemicals and radiation that damage DNA), many cells (both normal and cancerous) undergo apoptosis because of the vast amounts of unrepaired DNA damage. Those cells that do not directly undergo destruction are temporarily arrested in the cell cycle to allow for DNA repair. Many of these survivors are cancer cells since they evolve the ability to become unresponsive to apoptotic signaling.

While we often think of cancer as a disease of excessive cellular replication, it is essential to note that cancer is also a disease of cells that fail to die when they should—cells that are, to turn a phrase, "hard to kill." By failing to abide by the innate signaling mechanisms that root out damaged cells, cancer cells threaten the survival of the whole organism of which they were once an integral part.

Some survivors enter the state of cellular senescence, also called *replicative exhaustion*, in which, despite their quiescent state, they play a significant role in shaping the landscape of tumor evolution. Though non-proliferative, these cells secrete cytokines, growth factors, and other signaling proteins that promote the cellular phenotypes required for the repair of the damaged tissues.[254]

These signals—which inhibit apoptotic signaling, enhance the proliferation of new cells, reduce cellular adhesion to allow for cell movement, and enhance the activity of matrix metalloproteinases—establish optimal conditions for the repair and remodeling of damaged tissues. In the tumor microenvironment, such signaling promotes tumor growth and invasion, as the surviving cancer cells respond by developing the characteristics of cells actively engaged in wound repair, phenotypes also found in invasive cancer cells.[255]

Genotoxic treatment increases genetic instability in cancer cells, as the treatment directly causes additional mutations and genetic rearrangements with associated damage to caretaker and gatekeeper functions. As a result, the mutation rates in the cancer cells increase, and the mitotic machinery is rearrangements, and changes in chromosome copy number—thereby providing for increased diversity amongst the surviving clones.

In the face of the challenges of the tumor microenvironment, this combination of cellular diversity and genetic instability enhances the selective pressure on the cancer cell clones. This dynamic leads, in time, to the demise of most of the cancer cells, with a small percentage of survivors. These are the cells that have adapted over time to the changing conditions of the TME. These survivors are the most virulent cancer cell clones that can forge ahead under intense selective pressure in the TME as they strive to outcompete the other clones for available resources.

Ironically, at the tip of the double-edged sword, we arrive at the realization that the therapy itself provides for tumor cell progression, as "treatment-induced damage to the microenvironment can promote a chemoresistant niche of residual disease that subsequently serves as the nidus for relapse."[256] Nidus, from the Latin meaning "nest," refers to the site of origin of a disease. Simply stated, genotoxic treatment may kill cancer cells, but it also serves as a potent driver of cancer cell evolution.

Considering the stem cell hypothesis, we can consider the cancer cell as either a normal stem cell gone rogue or, perhaps, a cancerous differentiated cell that has gone "backward" on the developmental path to become a cancer stem cell. We therefore cannot prove that cancer is caused by a renegade stem cell.

We can say that the cells capable of transplanting a viable tumor into immunologically compatible hosts are the only cells that can seed new tumors. This is a tiny percentage of the cells in a tumor mass. Whether these cells are bona fide stem cells or differentiated cells that have "de-differentiated" to a stem-like state or some combination thereof, we do know these cells have many characteristics in common with stem cells. Regardless, we will consider the term "cancer stem cell" as a placeholder for a cell of unknown origin that appears to be the driver of many (perhaps most) types of cancer.

Cancer stem cells share many phenotypic characteristics with normal stem cells. Both types of cells mature within defined niches responsive to local conditions: cancer cells within their niches in the tumor microenvironment and stem cells in specialized compartments called *stem cell niches* within the micro-architecture of the organs. Both types of cells are quiescent, dividing infrequently and capable of producing more stem cells as well as differentiated, more rapidly dividing cells.

Both types of stem cells are long-lived as well as resistant to apoptosis and the effects of toxic agents. Both normal and cancer stem cells have elevated levels of ABC transporter proteins in their cell membranes. This characteristic provides the precious stem cells, the pluripotent reservoirs of future generations of cells, with an extra layer of protection from the environment, thereby improving their fitness so they can fulfill their critical roles for future generations.[257]

The cancer stem cell hypothesis provides a compelling explanation for why cancer is so hard to defeat. Resident in their micro-niches, where they are protected (at least in part) from therapeutic intervention and stimulated to adopt mesenchymal phenotypes by the stromal cells in the TME, cancer stem cells continue to evolve. Over time, the evolving cells can detach from the tumor, invade adjacent tissue, and move through the stroma toward the circulation to begin their long and arduous journey in search of a new home.

Biological agents such as antibodies and other protein-based therapeutics have altered the cancer treatment paradigm from radiation and the use of drugs that poison cancer cells with genotoxic chemicals to modalities that modulate the biochemical circuitry using highly specific biological molecules.

Drugs such as Herceptin and Rituxan—which are used for breast cancer and B cell malignancies such as leukemia and lymphoma, respectively—have provided tremendous benefits to cancer patients for over two decades. For some patients, treatment halts tumor growth (this is called *progression-free survival, or PFS*), while others derive a survival benefit, an increase in *overall survival* (*OS*). The two outcomes do not necessarily correlate, such that an increase in PFS does not always result in an increase in lifespan compared to untreated patients.

Despite these advances, biologicals are nonetheless limited in their ability to provide long-term remissions over decades. The mechanisms responsible for these limitations are like those found in the case of genotoxic agents. In both cases, the target molecule can mutate so the therapeutic no longer recognizes its target. In addition, as noted previously, cancer cells can evolve an alternative biochemical pathway to circumvent the one blocked by the therapeutic agent.

Given the limitations of both genotoxic and targeted biological therapeutics, is there a better way to achieve long-term results in cancer treatment? The answer to this question will emerge in the following chapters.

Chapter 12

Targeting Tumors the 21st Century Way

On rare occasions, tumors disappear for no apparent reason. This phenomenon of spontaneous tumor regression—also known as *spontaneous remission*—has perplexed physicians and scientists for centuries.

Documented cases of spontaneous remission of cancer go back to the thirteenth century, when an Italian priest named Peregrine Laziosi had an extraordinary experience. Laziosi was suffering from a large tumorous mass in his lower leg that later erupted into a foul-smelling ulcer. The ulcer emitted a "stench that was so over-powering that friends who were taking care of him could only remain nearby for a limited amount of time."[258]

Following the appearance of the ulcer, the tumor began to shrink. The priest eventually recovered, living to the age of 85.[259] The best conceivable explanation at the time was that a miracle had occurred. They did not, and could not, know that the severe infection stimulated the suffering priest's immune system, which, in turn, attacked his tumor. That idea wouldn't come of age until the twentieth century.

There are also reports from antiquity that say physicians had resolved cancers using procedures causing a fulminant infection at the tumor site. A papyrus scroll discovered in the sixteenth century on the Iberian Peninsula (modern Spain and Portugal) described the use of a poultice for the treatment of tumors during the time of the famous Egyptian physician Imhotep (around 2600 B.C.E.).[260] Naturally, the Egyptians could not have envisioned that the resulting infection from the poultice—a collection of herbs and leaves that may not have been filthy, but was far from sterile—had infected the patients and boosted their immune responses. Sometimes the patients recovered after treatment. While no further information exists, it is likely that many of these patients succumbed to cancer even though they too had developed infections.

With the Enlightenment in the late eighteenth century came a new appreciation that science plays a key role in natural phenomena (perhaps as important, some dared claim, as the Creator). Recurrent observations that cases of spontaneous tumor regression were often associated with the presence of a high fever and active infection encouraged the use of bacteria in treating human tumors. Limited options were available beyond surgery and prayer.

The most famous and influential physician who investigated the link between tumor regression and infection was William Bradley Coley. Born in Connecticut during the second year of the Civil War and educated at Yale University and Harvard Medical School, Coley was a gifted young surgeon at the New York Hospital in 1890 when a young woman came to his clinic who was suffering from a painful sarcoma. She had a malignant bone cancer in her hand, and despite his best surgical efforts to save the patient, the hand's amputation was an inadequate treatment, as the sarcoma had already metastasized to her liver and lungs.[261,262]

Less than a half-year later, on January 23, 1891, his patient, Elizabeth Dashiell, died at the age of 17. Coley was determined to delve deeper into the nature of the sarcoma that had taken Elizabeth's life. He dove into the hospital's archives and stumbled upon a case that he thought might provide some insight.

Fred Stein was a German immigrant who was successfully treated at The New York Hospital in 1883 for a persistent neck tumor that kept growing back after repeated surgical excisions. Following his final surgery, Stein developed an infection with *Streptococcus pyrogenes*. *Streptococcus pyrogenes* is a common and potentially deadly pathogen that leads to an excruciating condition called Saint Anthony's Fire in recognition of the searing pain it inflicts on its victims.

The physicians at New York Hospital expected that Mr. Stein would die in short order, but the patient far exceeded medical expectations. Over the next few weeks, his tumor shrank and then disappeared. Fred Stein was discharged from New York Hospital with a nasty scar on his neck to remind him of the tumor that had made his life a living hell.

Coley wondered what had happened to Mr. Stein. Had the tumor recurred? Was he still alive now, eight years hence? Unfortunately, there was no follow-up information in the records.

While William Coley was a trained surgeon, he also had the instincts of a sleuth. Like any good scientist provided with a lead that might help get to an understanding, he could not let go of this thread; he needed to see where it might take him. To do so, he had to do something a tad unconventional for a renowned man of medicine: he visited the German tenements on Manhattan's Lower East Side.

As he searched the slums for information, Coley was fully aware that he faced a challenging task in determining the whereabouts of a man named Fred Stein in a German community. To compound the challenge, Coley didn't know anything about Mr. Stein's activities since leaving the hospital eight years ago.

Coley's determination and persistence paid off. He not only uncovered information about Fred Stein, but the physician also found the cancer survivor himself, alive and well. This was truly astonishing. Coley realized he had found evidence supporting the hypothesis that infection at the site of a tumor can trigger regression of the cancer. Infection had truly ignited a biochemical fire under Mr. Stein's immune system.

Coley took this line of reasoning one step further. He instantly hit upon an intriguing idea. If the underlying hypothesis was true, he might be able to duplicate the conditions that lead (at least in some patients) to tumor regression by inducing fever and infection in cancer patients. It might be possible to "kick-start" the immune system so it could recognize and attack the tumor.

This was an avenue of inquiry that begged resolution. As a surgeon in the bone cancer unit at New York Hospital, Coley saw many patients with malignant cancers who had little hope for survival. The surgeon had at hand an ideal test population for evaluating his hypothesis. There was little downside in these investigations, as the patients could not be saved using surgical techniques.

Creating an infection with a dangerous microbial agent while keeping the patients alive presented a tricky balancing act. Too extreme an immune response to the toxin had the potential to kill the patient; the failure to achieve an adequate immune response to resolve the tumor only subjected the patient to a tortuous treatment without benefit.

Today, no physician would dare think it was acceptable to experiment on patients, even when treating a fatal disease. But Coley practiced medicine in the days before regulatory requirements mandating clinical trials that demonstrate the treatment's safety and efficacy. It was, therefore, a simple matter—and in Coley's eyes, an act of compassion—to offer patients the chance to throw the dice on his experimental treatment.

Coley began cautiously, slowly building up the potency of his therapeutic preparations until he found one that seemed to work: a mixture of the dreaded *Streptococcus pyrogenes* with a bacteria called *Serratia marcescens*, a mildly pathogenic bacterial species that generally only infects people suffering from an immune deficiency.[263] This combination of bacteria, dubbed "*Coley's Toxin*," appeared to be effective in some patients.

The treatment was particularly effective in patients with sarcomas. These cancers are rarer than carcinomas, which comprise the vast majority of cancer cases. Since Coley was a bone specialist, most of the patients in his clinic were suffering from sarcomas, the most common variety of bone cancer. He therefore had a large patient population for his studies.

Though his results were encouraging, Coley met vociferous resistance from his colleagues. Without knowledge about human immunity, many could not even fathom the scientific value of Coley's experiments. At the time, the medical establishment used surgery and a handful of largely ineffective drugs and tinctures to treat cancer. The idea that patients could somehow heal themselves from within seemed nothing short of quackery.

On top of that consideration, the very idea of treating sick patients with infectious agents was difficult for physicians, who had pledged to "do no harm," to accept. Furthermore, the therapeutic agent was inconsistent in performance, as its preparation included a critical heat treatment designed to dampen (but not severely deplete) the infectivity of the bacterial population. A great deal of skill and learned technique was needed to create an effective batch. It was difficult for most practitioners to produce material that was sufficiently dampened in infectivity to avoid killing the patient while retaining the ability to generate a potent immune response.

With respect to patient safety, the treatment was, in fact, extremely dangerous, excruciating, and potentially life-threatening. It resulted in high fever, unstoppable chills, and searing pain at the injection site. Injections of increasing strength were given over several months, with injections every other day. This was a long, arduous, and debilitating process for the patients. In some instances, the treatment hastened the patient's death. However, the fatality rate of the patients in Coley's original cohort was only 6 out of 1000, an extraordinarily low fatality rate (0.6%) for the treatment of advanced cancer, even by today's standards.

The advent of chemotherapy and radiation in the early twentieth century focused attention on these exciting new therapies. These new therapeutics were far more standardized than Coley's Toxin as well as compatible with the tenets of traditional medical practice. The American Medical Association panned Coley's treatment in 1894, only three years after reports of his first clinical successes.[264] The AMA alleged, "There is no longer much question of the failure of the toxin injections, as a cure for sarcomata and malignant growths. During the last six months the alleged remedy has been faithfully tried by many surgeons, but so far not a single well-authenticated case of recovery has been reported."[265]

This did not stop the pharmaceutical company Parke Davis from developing one of Coley's preparations as a prescription drug that went on the market in 1899 and was subsequently used by some practitioners for treating incurable metastatic cancers. Ongoing reports of the efficacy of the treatment led to a reversal on the part of the FDA in 1936, when the agency charged with the safety and efficacy of the U.S. drug supply stated, "Its use in inoperable cases may be quite justified."[266] Coley's Toxin remained on the market until 1952, when Parke Davis decided the drug was not used in enough patients given the availability of the new chemical agents. The FDA finally ordered the cessation of the manufacturing of Coley's Toxin due to its known toxicity and the lack of appropriately controlled and documented clinical trials.

Despite the lack of data from clinical trials, there is data that we can use to assess the clinical impact of Coley's Toxin. After Coley's death in 1936, his daughter Helen began work on her famous father's biography. During her research, she stumbled upon a stack of files containing the records of the patients who were treated with the toxin by Coley and his colleagues at New York Hospital.

Like her father, Helen developed the talents of a detective. She tracked down as many of the Coley's Toxin patients as possible to determine what had happened to these cancer sufferers. Helen's research became a long and arduous quest, and she did not publish her findings until 1953.

The results were astonishing. Out of a total of 1200 cases, 270 of the patients eventually recovered and did not develop the sarcoma again.[267] The achievement of lifetime remissions ("cures") of advanced cancer in 22.5% of the treated population (almost one in four) would be considered a resounding clinical success even by today's standards for clinical outcomes in the treatment of malignant cancer.

The published findings on Coley's Toxin in the treatment of sarcomas stimulated further investigations by a small cadre of believers in the importance of immune stimulation in cancer treatment. One survey published in 1959 supported Helen Coley's findings, showing a 53% success rate in the treatment of 186 sarcoma patients using Coley's Toxin.[268]

It is important to note that by success, this data does not refer to the 5- and 10-year survival metrics often used in benchmarking cancer therapies. Using this type of measurement, we cannot get a sense of the true patient experience, as this data does not specify how often the patient received follow-up treatment, how severe the side effects may have been, or even if cancer returned later in life.

In the case of the sarcoma data from Helen Coley's analysis and the 1959 study reported above, the success metric was a total remission. This means that all patients who were counted as successfully treated were free of the sarcoma throughout the remainder of their lifespans (or, alternatively, during the entire follow-up period, as some patients were still alive at the time of data collection). One of the successful patients reported in the 1959 study had remained sarcoma-free for 62 years.

Even amongst the patients who did not have a successful outcome in the 1959 study (47%), many responded to the treatment and enjoyed a long remission that was followed, years later, by tumor recurrence.[269]

The demonstration of the clinical success of Coley's Toxin rested on the results with sarcomas. Additional data provided by adherents of Coley's approach showed that his method had promise for the more common carcinomas, including carcinomas of the breast, ovary, uterus, cervix, kidney, colon, and skin.[270] As a final testament to the anti-tumor activity of Coley's Toxin, a comparison of the results for patients treated with Coley's method with the National Cancer Institute's data for patients diagnosed in the early 1980s shows comparable (and sometimes improved) 10-year survival for Coley's method in three types of carcinomas: kidney cancer, ovarian cancer, and breast cancer.[271]

This is a truly startling story demonstrating that the medical community, and the FDA, were too harsh in their criticism of Coley's treatment. Unconventional? Certainly, but the evidence shows that William Coley was on target in his dedication to the idea that the stimulation of the immune system can have significant therapeutic effects in cancer patients. For the patients who recovered from their inoperable cancers, the outcome of Coley's treatment must have seemed like magic.

Paul Ehrlich was an early and strong proponent of the idea that stimulation of the immune system might exert a potent therapeutic effect in cancer patients. In 1909, he noted, "Cancer would be quite common in long-lived organisms if not for the protective effects of immunity."[272]

Early investigations of tumor immunity in Ehrlich's time involved the use of extracts from the primary tumors of metastatic cancer patients to stimulate an anti-tumor immune response in them.[273] In these experiments—performed in the first two decades of the twentieth century—suspensions of the patient's tumor cells were readministered to the same patient. This type of procedure, in which the patient's own cells are reintroduced, is called an *autologous cell transplant* (*auto* is Greek for "self").

Patient responses in these attempts to generate immunity to the tumor were negligible. In the few cases where there was some evidence of tumor regression, these responses were short-lived; tumor growth was observed within weeks to a few months. Most researchers concluded that immunization of patients with tumor-derived material was not a winning approach to cancer treatment.

Other researchers were trying a different approach for stimulating anti-tumor immunity. The rabies virus was under investigation early in the twentieth century as a potential immunostimulatory agent in cervical cancer patients. In these treatments, live rabies virus was inactivated using chemical agents and heat. The weakened (*attenuated*) virus preparation was unable to infect host cells and thereby deemed acceptable for use as a therapeutic agent.[274]

While a short-lived shrinkage of the tumor was sometimes observed, these preparations did not demonstrate the potential to generate a significant anti-tumor response. There were significant side effects, including, on occasion, a deadly viral infection from the therapeutic agent itself. In these cases, the preparation had not been sufficiently inactivated to eliminate its infectivity. As a result, the promise of viral therapies for cancer suffered a significant setback.[275, 276]

Other early attempts in cancer immunotherapy involved the use of anti-tumor antibodies raised in animals. In these experiments, animals were immunized with extracts from the patient's tumor cells. Next, the serum containing the antibodies created by the animal's B cells was administered to the patient who supplied the tumor tissue used for immunization of the animals.

The use of antibodies generated outside the patient as therapeutic agents is known as *passive immunotherapy*. Passive immunity lasts only as long as the immune system is exposed to the injected antibodies, which are in fact the therapeutic agents. *Active immunotherapy* involves the creation of a long-term response by stimulating the recipient's own immune system to create the antibodies (and/or generate long-term T cell memory). For example, the measles vaccine is a form of active immunotherapy that generates an immune response against the measles virus. Such a response can last for decades.

As in the case of active immunotherapy with tumor extracts, administration of the animal antibodies (generally derived from rabbits, sheep, horses, dogs, donkeys, or goats) generated, in some patients, objective responses in the form of tumor regression. The regression was of a similar short-lived variety as that observed with the tumor extracts. In addition, since the animal serum itself contains many nonhuman proteins, it generates significant side effects—most notably, an immunological condition known as *serum sickness*—due to the patient's immune response to the foreign antigens in the animal serum.

Unfortunately, this host response—the formation of host antibodies—includes the formation of human antibodies to the animal-derived anti-tumor antibodies. When the patient creates antibodies against the animal antibodies (an *anti-antibody*, and yes, it can be confusing), the animal antibodies are bound by human antibodies and thereby inactivated before they can reach their target. It didn't take long to recognize that this approach was doomed to failure.

By the 1960s, the scientific underpinnings of human immunity were sufficiently established to suggest another approach that leveraged the anti-tumor immune response in patients. Perhaps the administration of blood from patients who had undergone spontaneous tumor regression would be helpful as a therapeutic agent. The blood from such patients should contain, the reasoning went, immune cells that had already achieved a potent anti-tumor response. Could such immune cells be used in other cancer patients?

Trials in cancer patients showed limited success, with regressions observed in only a few melanoma patients. The promise of this approach was vastly augmented by the recognition that the blood contains specific types of immune cells responsible for attacking tumors. By isolating the lymphoid cells (cells of the adaptive immune response, the B cells and T cells) from the circulation, it was possible to administer—to either the original patient or to other patients with a similar cancer type—the B and T cell populations from a patient who had responded to the tumor.

Once again, occasional regressions were observed. While this approach showed some promise, it did not deliver results suggestive of an effective cancer therapy. There was something missing in the design of effective immunotherapies capable of generating a potent and lasting anti-tumor response.

The fact that tumors advance in the body demonstrates that the human immune system is sometimes unable to recognize and destroy them. Conversely, the fact that human cancers sometimes undergo spontaneous remission strongly suggests that the immune system can detect and destroy cancer cells, even, on occasion, when the cancer is in an advanced state. A logical corollary of these observations follows: if we could understand the mechanisms that allow tumors to avoid immune surveillance and thereby resist the natural immune response, it might be possible to intervene therapeutically to awaken the senescent immune system during tumor progression.

Sounds reasonable enough. But the secret to training the immune system to attack tumors presented a formidable scientific challenge. There were, however, some clues from another areas of medicine: organ transplantation. The discovery of the cell-killing activity of T cells was not made in the context of cancer. Rather, it was observed that T lymphocytes participate in the rejection of grafted tissues and organs (a phenomenon called *graft versus host disease*). T cells are responsible for targeting and destroying cells that display foreign ("non-self") antigens to the host immune system.[277] This discovery suggested that T cells might also be responsible for killing tumor cells in the body.

The trick, it seemed, was in figuring out how to create the conditions for the generation of a large population of T cells with anti-tumor properties. The search for the answer to this problem has consumed over half a century of intensive research.

Steven A. Rosenberg was born in the summer of 1940 to Orthodox Jewish immigrants from Poland who settled in the Bronx. Educated at the prestigious Bronx High School of Science, followed by Johns Hopkins University and Harvard, Rosenberg established his future path at an early age. "As soon as I stopped wanting to be a cowboy at the age of 5, I knew I wanted to be a physician—and not just a physician, but a medical researcher, as well."[278]

As a young physician, Rosenberg witnessed an apparent miracle: the full recovery of a young man with metastatic cancer without any therapeutic intervention. This spontaneous remission of advanced cancer stunned young Dr. Rosenberg so much that he decided there and then on his life's mission as a physician. He would figure out how such remissions occur and use that knowledge to treat patients whose cancers had advanced far beyond the current ability of medical science to intervene.

As he searched for an explanation for this miracle, Rosenberg realized that the liquidation of an advanced cancer must be through the action of the immune system, which has the innate ability to heal the body from the ravages of disease. He also recognized that in most cancer patients, who do not experience the grand fortune of a spontaneous remission, the immune system must be in a state where the cancer prevents it from recognizing, locating, and destroying the tumor cells.

Taking this line of inquiry one step further, he came to a startling conclusion. If one could overcome the barriers presented by the presence of the cancer to the intrinsic power of immunity, it might be possible to unleash the immune system on the cancer, thereby increasing the likelihood of a remission. In the best case, it was conceivable the immune system's full potential could be realized, resulting in the elimination of the cancer cell population from the body.

Rosenberg set his sights on pursuing this remarkable vision. At the time of this writing, he has been at the NCI working toward that goal for 50 years. For this reason, he "is widely considered the father of cancer immunotherapy,"[279] with an extensive list of astounding accomplishments that have advanced the treatment of cancer on multiple fronts.

Early in Rosenberg's career, there was scant evidence supporting his vision. While it had long been hypothesized that the immune system must provide a protective barrier against cancer, little direct experimental evidence supported this contention. Like Paul Ehrlich before him, Rosenberg believed in the potential of immunity to vanquish cancer. After all, he had seen it happen right before his eyes.

Animal studies designed to determine whether T cells can recognize and destroy cancer cells were initiated in the mid-1960s. There was an inherent problem in this work. Like healthy human cells obtained directly from the body, T cells do not grow well (or survive beyond several days) in cell culture. It was therefore not possible to obtain large numbers of T cells using those found in an animal's tumor as a source.[280]

There was, however, another way to generate significant numbers of antigen-stimulated T cells. After immunizing a genetically similar animal (a *syngeneic* recipient) with the antigens derived from an individual tumor, the T cell population (including the antigen-stimulated T cells) can be recovered from the spleen of the immunized animal and injected into a tumor-bearing syngeneic animal (including the original tumor-bearing animal itself). The close genetic match between the animals reduces the likelihood of an immune response to the injected cells.

The results demonstrated the merit of the hypothesis. The treatment was able to shrink tumors in animal studies performed by several investigators.[281] In one study at the University of Washington School of Medicine, the response rate, in terms of complete tumor regression, was in the range of 35-40%.

The regressions were characterized by the presence of significant numbers of T cells in the tumor bed, demonstrating the importance of an immune infiltrate in the regression event. Finally, a rather prescient statement pointing to the importance of using large numbers of T cells in immunotherapy, and treating the tumor as early as possible, appears in a half-century-old report in the journal *Cancer Research*. "A greater frequency of cures was obtained by treating smaller, though still palpable, tumors with larger numbers of spleen cells."[282]

In 1976, the discovery of an immunostimulatory protein called *T Cell Growth Factor* (*TCGF*) completely changed the strategy of immunotherapy by introducing an agent that could be used in cell culture to stimulate the proliferation of T cells. This immunomodulatory protein (a cytokine)—discovered by Doris Morgan, Francis Ruscetti, and Robert Gallo—was capable of stimulating T cells to reproduce in cell culture.[283] Later re-named *Interleukin-2* (*IL-2*), T Cell Growth Factor provided researchers with a tool for generating large populations of antigen-stimulated T cells.

Knowing that the place to find the T cells of interest was in the region of the tumor, Rosenberg and his team were quick to jump on the opportunity provided by the discovery of IL-2. The T cells found in the tumor bed, known as *tumor-infiltrating lymphocytes* (*TILs*), were "just what the doctor ordered." Their presence in the vicinity of the tumor suggested they might be responsive to tumor cell antigens.

These T cells could be taken from the tumor, grown in cell culture media containing IL-2 in the laboratory, and then infused into a syngeneic recipient animal. The first of these studies from Rosenberg's lab at the NCI "in 1982 demonstrated that intravenous injection of immune lymphocytes expanded in IL-2 could effectively treat bulky subcutaneous ... lymphomas."[284]

Rosenberg and his team at the NCI—where Rosenberg was appointed Chief of the Surgery Branch on the last day of his residency at the tender age of 34—also realized that IL-2 had the potential to serve as a chemotherapeutic agent by stimulating the production of T cells inside patients. Activation of the subset of T cells known to immunologists as *CD8-plus (CD8⁺) cytotoxic T cells* might help battle cancer because CD8⁺ T cells can kill both pathogens and tumor cells. They achieve this by punching a hole in the target cell membrane (using a protein called *perforin*) and releasing *granzymes*, enzymes that degrade proteins and induce apoptosis in the targeted cells.

The path forward to evaluate the hypothesis that IL-2 could be used to expand the anti-tumor T cell population was fraught with frustration. The NCI team persevered. After proof of principle was obtained in mice, clinical trials in which patients were injected with IL-2 or IL-2-treated immune cells obtained from the patient provided a nice round number of positive results in the first 66 patients: 0.[285]

For the 67th patient (in 1984), a new treatment regimen with a higher dose of IL-2 was given to a Navy officer named Linda Taylor, who was suffering from metastatic melanoma. Amazingly, her tumor disappeared, and she remained tumor-free for over three decades. This astounding outcome demonstrated, for the first time, that it is possible to successfully treat human tumors with an immunomodulating protein in the absence of any therapeutic intervention directed at the cancer cells.[286] Further work showed that human tumors could be reduced in size by the addition of high doses of IL-2. Consequently, the cytokine was licensed as a therapeutic agent for renal cancer in 1992 and for melanoma in 1998.[287]

These pioneering studies on the use of IL-2 as a T cell-stimulating factor provided fuel for a revolution in cancer immunotherapy that birthed a new field of immunotherapy called *Adoptive Cell Transfer (ACT)*. The core idea of ACT is that living T cells can be removed from the body, treated in the laboratory to enhance their anti-tumor properties, and then infused back into the patient to generate a strong and specific immune response to the tumor. In this way, cancer can be treated with activated T cells, a living drug capable of reproducing inside the patient and leveraging the innate power of cytotoxic T cells to eradicate tumors.

To generate a strong anti-tumor response, the immune system must differentiate the tumor cell and its protein antigens from those found in a normal cell. In tumor cells, such foreign antigens might be mutated forms of "self" proteins, over-expressed normal proteins, and/or proteins expressed in cell types where they should not be present (or are present at inappropriate times during the lifespan). Furthermore, to mount an effective immune response, the innate and adaptive arms must work in concert to identify foreign antigens and stimulate naive (i.e., unexposed to antigen) B and T cells so a potent and orchestrated response can be generated to overcome the threat.

Starting in the late 1950s, research teams led by Richmond T. Prehn (National Cancer Institute), and George Klein (Karolinska Institute, Stockholm, Sweden) provided convincing experimental evidence for immunity's involvement in resistance to cancer. Using chemical carcinogens to induce sarcomas in mice, they established actively growing cancers in the test animals. After the tumors were established, the animals were resistant to further challenges when samples of their tumors were reimplanted elsewhere in their bodies. However, when the tumor-bearing mice were implanted with the same type of tumor from a different animal, the implanted sarcomas grew rapidly.[288]

To minimize the immune response in the form of tissue rejection, these experiments used syngeneic mice from the same inbred colony. In these animals—in the presence of low levels of genetic variability (there is always some)—an immune response was likely to be generated by a tumor-driven change in a protein sequence recognized as "foreign" by the host immune system. The result would be a graft-versus-host reaction leading to rejection of the tumor.

The experimental evidence "clearly showed that tumors harbored one or more specific antigens capable of eliciting immunity and long-term immune memory." According to Richmond Prehn, "Because members of an inbred strain were assumed to be genetically and thus antigenically identical, it was concluded that the immunity must necessarily be directed against the tumor tissue *per se.*"[289]

Prehn and Klein performed their work in an era when we did not understand how T cells recognize antigens, nor did the technology exist to directly identify the molecular target of a T cell. It would take a quarter century to understand how T cells recognize their target antigens. It would take another three decades to discover unambiguous evidence of a human tumor-specific antigen.

Exactly how the immune system primes its cellular network to discriminate foreign from self-antigens—and the basis of antigen-specific responses by the adaptive arm of immunity—was a problem that drove an intensive research effort throughout the 1970s. The discoveries about the fundamental nature of adaptive immunity were enabled by emerging technologies wrought by the age of recombinant genetic engineering, with vast improvements in the ability to insert genes of choice into cells grown in laboratory cultures.

These discoveries were accompanied by biochemical advancements that allow for isolation and detailed structural analysis of the protein product encoded by the gene of interest. As the technology advanced, the explanation of how T cells recognize antigens was finally revealed.

As an incoming postdoctoral researcher in the laboratory of Stuart Schlossman at the Dana-Farber Cancer Institute, Ellis Reinherz arrived with an inquisitive mind and well-formed scientific ambition. Like Steven Rosenberg, Reinherz was "motivated by a clinical observation and desire to understand its basis."[290]

The observation that he found fascinating—the problem that he could not let pass by—related to the existing data on the clinical management of pediatric *acute lymphoblastic leukemia* (*ALL*), the most common form of childhood leukemia. Pediatric ALL is fast-moving and perilous. Yet, newly developed chemotherapy regimens were remarkably successful, achieving remission rates in the range of about 80%.[291]

This raised an interesting question: was there simply a random 20% of patients who failed to respond or, alternatively, was there something special about 20% of the patients for whom the treatment cannot be effective—such that they are doomed regardless?

Reinherz believed it to be the latter. There had to be a reason some recipients were more resistant to the treatment than others. He believed the key to the puzzle had to reside in the adaptive immune system, amongst the T and/or B cells.

The initial presentation of Reinherz's research interest to his clinical mentor around Christmas time in 1977 did not go well: "He looked at me as if I had three heads and simply commented that those who die have 'poor protoplasm'."[292]

Perhaps the young scientist's idea was ahead of its time, and his mentor was not ready to embrace it. What Reinherz presented may have seemed to others a long-shot proposal based on suspicion rather than evidence. He obviously had yet to establish a reputation at the level of the more seasoned scientists around him. Yet the idea was too intriguing to let go. At appreciable risk to his young scientific career, Ellis Reinherz was ready to forge ahead on his own.

This seems an appropriate time to note that scientific breakthroughs cannot be found within the existing base of knowledge in the field. As Thomas Kuhn described in his book, *The Structure of Scientific Revolutions*, the "normal science" that works inside the paradigms of the times is certainly necessary to move science forward, but true leaps forward—the revolutionary changes that alter the trajectory of a field forever—come about when scientists step outside of the current paradigm and invent a new one.

Technological advancements can be a driving force for a scientific revolution. The invention of PCR provides an example. Without PCR, we could not amplify DNA to obtain enough for accurate sequencing, and we therefore could not have determined the sequence of the human genome.

Other scientific revolutions happen through the power of the mind. The most obvious example is Albert Einstein, whose uncanny ability to "ride on the tip of a light beam" and conjure other fantastical "thought experiments" provided the foundational understandings of space, time, gravity, matter, and energy at the core of modern physics and cosmology.

Determined to pursue the answer, Reinherz left his clinical hematology fellowship and joined the laboratory of Stuart Schlossman in Boston. Dr. Schlossman was Chief of the Division of Tumor Immunology at the renowned Dana-Farber Cancer Institute. Reinherz chose to join the staff of that laboratory because they were engaged in research on the functions of the cells of adaptive immunity.

The Division Chief was happy to bring the eager young scientist on board. His lab used the hybridoma method of Kohler and Milstein, published only two years earlier, to generate antibodies against immune cells, particularly T cell populations obtained from human donors. In 1979, Reinherz found a protein on the surface of human T cells called *cluster of differentiation 4* (*CD4*), which was present on about two-thirds of mature circulating T cells.

He found that these cells demonstrate immunological "helper" activity. When stimulated by antigens, these cells secrete cytokines that orchestrate the delicate balance between the cells of adaptive and innate immunity. Shortly thereafter, T cells with *cytotoxic* (cell-killing) activity that have a protein called *cluster of differentiation 8* (*CD8*) on their surface were discovered.[293]

It is important to note that these discoveries about CD4 and CD8 just barely predated the emergence of a strange and terrifying viral ailment first called Gay-Related Immunodeficiency Disease (GRID) in the early 1980s. Later, GRID was more appropriately named *Acquired Immunodeficiency Syndrome* (*AIDS*). AIDS researchers reaped the rewards of the raft of immunological discoveries in the previous two decades, perhaps most significantly by the CD4 discovery. CD4 is the protein used by the HIV virus to latch onto T helper cells during infection.

Using their panel of monoclonal antibodies to diverse T cell populations in various tissues, Reinherz and his collaborators found something unexpected in the T cell samples derived from lymph nodes. There was a significant population of cells that expressed both CD4 and CD8. They found that these "double-positive" cells were the precursors of the mature lymphocytes found in the circulation and inside tissues that express either the CD4 or the CD8 marker, but not both.[294]

In cell populations derived from the thymus—a small gland in the chest that is the source of immature T cells—they found a small population of T cells with neither marker. Amazingly, most of the pediatric ALL patients who failed to respond to the chemotherapeutic regimen had an observable population of these *"double-negative" T cells*, while the responders did not.[295]

Ellis Reinherz had his answer. He was right; some patients had a defect in their T cell populations that was inhibiting T cell maturation and dooming them to therapeutic failure. This was a significant finding that explained, at least in part, the molecular basis of the poorer response in these patients.

In 1980, Reinherz's research productivity prompted his promotion to Assistant Professor at Harvard Medical School, with his own laboratory dedicated to "defining T cell antigen recognition, including the identification of the *T cell receptor*."[296] By T cell receptor, Reinherz was referring to the molecular structure on the T cell surface that interacts with antigens. At the time, the identification of this receptor was a highly competitive field of research. It was evident that understanding the molecular basis of T cell antigen recognition would provide significant insight into how T cells perform their biological magic.

Once again, Dr. Reinherz was challenged by higher authority. The mission statement for his new laboratory was criticized by Dana-Farber President and Nobel Prize winner Baruj Benacerraf, who shared the prize in 1980 with Jean Dausset and George Snell for discovering the Major Histocompatibility Complex. Concerned about the young scientist's stated research objective, Benacerraf wondered aloud why Reinherz believed he could succeed in a field already brimming with more experienced researchers who were also seeking the molecular secrets about how T cells interact with antigens.

Reinherz took the feedback in stride and moved on despite the Nobel laureate's concern for his fledgling career as a new faculty member. The nature of the molecular basis of antigen recognition by T cells had occupied a prominent position in the minds of immunologists and biochemists for decades, including his. Someone had to figure it out. Why not him?

Using the panel of T cell-reactive monoclonal antibodies available in his lab, Reinherz and his colleagues discovered, in 1980, that an antibody directed against a protein found on all mature (fully differentiated) T cells called *CD3* was capable of binding to and shutting down T cell function.[297] By blocking the activity of CD3, the ability of T cells to recognize their antigens was obliterated. CD3 was somehow involved in antigen recognition. Using other antibodies raised against T cells, they showed that CD3 was associated with two other proteins embedded in the cell membrane of T cells. These proteins had molecular weights of about 49 and 43 kilodaltons. The 49 and 43 kilodalton protein chains—named α and β—are linked together to form a dimer comprised of the two different protein chains (this is called a *heterodimer*). These three proteins, CD3 and the heterodimer, comprise the T cell receptor—the protein complex on the T cell that recognizes its antigen.[298]

In a manner analogous to the variable sequences in antibodies, the heterodimer contains variable regions in the sequences of its two constituent chains that provide for unique specificities amongst the T cells. By means of complex biochemical interactions between the constituents of the T cell receptor, the antigen, and the HLA proteins on antigen-presenting cells to which the antigen is bound (and presented to T cells), the T cell response is able to cope with the millions upon millions of antigens encountered by the immune system over the lifespan.

The search for tumor-specific antigens in mice initiated in the 1950s by Prehn and Klein remains a highly active field of tumor immunology research. Throughout the history of the field, a fundamental question has lingered and remained at the core of our understanding of tumor immunity: what, exactly, is a tumor-specific antigen, and how can we recognize and exploit it for cancer immunotherapy?

There are cases where a protein normally expressed at low levels undergoes a vast increase in its expression level in a cancer cell. In such instances, the use of a therapeutic agent targeted at that antigen might be effective. However, since the target is shared by normal cells, there is a distinct possibility that such a strategy will cause significant side effects. This is because the agent acts not only on the targeted cancer cells, but also on normal cells that express the antigen on their surface. Therefore, such an antigen is, in a strict sense, not tumor-specific, but tumor-enriched.

A far superior targeting strategy would focus on a protein target that is only present in cancer cells—in cancer parlance, a *neoantigen* (*neos* is Greek for "new and recent"). Such an antigen can be a mutated form of a normally expressed protein or a protein that is not normally expressed in the tissue where the tumor is located. For example, tumors often express proteins that are normally found only in fetal tissues (these are called *oncofetal antigens*). Here again, as some tissues continue to express low levels of fetal antigens, targeting these proteins carries the risk of side effects due to interactions of the therapeutic with normal cells.

Targeting a unique protein—one that is biochemically distinct from the other proteins found in the normal cellular environment—is more likely to promote a specific immune response to cancer cells, with a presumably superior side effect profile due to the lack of interference with normal cellular processes. Clinical implementation of this enticing strategy suffers from one serious drawback. Finding such cancer-specific targets, while feasible, is a difficult feat, one that requires technological capabilities (described below) that have only recently been achieved.

Given the inherent power of the immune response, misdirection of adaptive immunity at "self" targets is a nasty affair that should be avoided at all costs. Herein lies the difficulty of overcoming transplant rejection and why tissue typing, which identifies the MHC gene profile of the cells in the sample, is so critical in establishing the compatibility, or lack thereof, between donor and patient.[299] The exquisite specificity and sensitivity to change of the immune system illuminate the risks in targeting a surface protein that is present, even in tiny amounts, on normal cells.

That said, while a novel cancer antigen found only on cancer cells is the ideal immunotherapeutic target, the daunting problem remains of identifying a handful of potential target antigen sequences from amongst the many millions of potential sequences found inside us. All the proteins that comprise us and that we are exposed to are presented to T cells in short peptides of about 5-7 amino acids. From the perspective of pure numbers, the search for tumor-specific antigens from amongst the millions upon millions of potential antigens screened by T cells during antigen presentation is an unfathomably complex task.

Fortunately, the science behind the effort to identify and target tumor-specific antigens has accelerated rapidly in recent years. This progress is due to the availability of sophisticated computer algorithms to interpret the massive data sets obtained from DNA sequencing and protein mass spectrometry.

Modern mass spectrometers can determine the amino acid sequence of vanishingly small amounts of a protein. This powerful technique for obtaining highly detailed information on the amino acid structure of proteins has developed at a substantial rate since the 1990s, with tenfold increases in sensitivity (ability to detect low levels) and resolution (ability to clearly identify one molecule from another) every four years or so.[300] As noted in a 2015 paper in *Science*, "Although a number of heroic studies provided early evidence for the immunogenicity of mutation-derived neoantigens, the technology to systematically analyze T cell reactivity against these antigens only became available recently."[301]

Studies in mice showed that most neoantigens contained mutations that failed to elicit strong T cell recognition when evaluated in T cell models in the laboratory that mimic the interactions between antigens and T cells. This finding suggests that once neoantigens are identified in each tumor, the pool of candidates contains only a few altered sequences likely to be recognized by T cells. In tumors with high mutational loads, such as melanoma and colon cancer, identifying T cell-reactive neoantigens requires efficient experimental strategies to focus the search on the antigens most likely to yield a positive response. For tumors with lower mutational loads—such as pancreatic and brain cancers—the problem of finding T cell-reactive neoantigens is compounded by several orders of magnitude.

When tumor-infiltrating lymphocytes were characterized in human tumors, researchers found that "only a very small fraction of the nonsynonymous mutations in expressed genes in these tumors leads to the formation of a neoantigen for which helper $CD4^+$ or cytotoxic $CD8^+$ T cell reactivity can be detected."[302] The phrase *nonsynonymous mutations* refers to those mutations that result in a change in the amino acid sequence coded by the DNA sequence; thus, *synonymous mutations* are those that alter the DNA sequence without altering the protein sequence, which can occur following a single DNA base change. This is because multiple codons are used for individual amino acids, such that a single change in a codon may not result in a change in the amino acid incorporated into that position on the polypeptide chain.[303]

Consequently, cancers with high mutational loads likely contain several nonsynonymous mutations that result in the presence of T cell-reactive neoantigens. In tumors with significantly lower mutational loads, there may not be any nonsynonymous mutations. Therefore, these tumors may lack T cell reactive neoantigens.

The biological mechanisms underlying the ability of T cells to recognize neoantigens include complex molecular interactions between the antigens and the HLA proteins on the surface of the antigen-presenting cells and the binding interactions between the antigen-HLA protein complexes and T cell receptors. Overlaying all this biological complexity in cancer is the impact of the local tumor microenvironment, which promotes immunosuppression.

Despite the complexity of these interactions, mathematical models that can predict the binding strength of neoantigens to T cell receptors while considering the specific HLA profile of the patient are beginning to show promise. Nonetheless, we have yet to achieve reliable predictive models in the search for tumor neoantigens that are likely to elicit a biological response from the host immune system.[304] If these neoantigens can be consistently identified in human tumors, the potential for successful immunotherapy and long-term remission becomes a realistic possibility for many patients, as we shall see in the next chapter.

Chapter 13

Living Drugs

The astounding therapeutic potential of the human T cell, so stunningly demonstrated by adoptive cell transfer, focused the attention of cancer researchers on applying the molecular tools of biochemistry and molecular biology to the genetic alteration of human T cells to enhance their ability to seek out and destroy living tumors. As Steven Rosenberg and his NCI team had shown, T cells can be isolated from a tumor patient's blood, genetically engineered in cell culture in the laboratory, and then reinfused into the patient as energized tumor-killing machines. This mode of treatment using the patient's T cells in an autologous transplant eliminates the risk of an immune response to T cells from a donor.

In an autologous transplant, the T cells taken from the cancer patient have been in the tumor's presence for months or even years. These T cells are primed by tumor antigens but unable to respond to their presence due to long-term antigen exposure. This constant exposure to antigens results in *T cell exhaustion*, whereby the T cell receptor cannot respond to the presentation of antigens that would, under homeostatic conditions, elicit a T cell response.

We can think of the exhaustion phenomenon as a desensitization of the receptor due to sensory overload. Receptor desensitization is a general feature of biological receptors. For example, opiate receptors that are continuously exposed to opiates for an extended period become less sensitive to the presence of the drug. As a result, the receptor's response diminishes over time.[305]

This reduction in sensitivity to opiates leads to dosage increases that only serve to desensitize the receptors even further, promoting intense cravings that are a hallmark of opiate addiction. This flexibility in response, where the receptor becomes desensitized, is another manifestation of biological plasticity. This capability provides a defense against receptor overstimulation, which can have dire consequences. For example, overstimulation of growth factor receptors contributes to oncogenesis in many cancers, including breast and colon cancers.

The adoptive cell transfer process pioneered by Rosenberg and his team at the NCI using tumor-infiltrating lymphocytes reenergizes the exhausted T cells that have infiltrated the tumor bed. It begins with a sample of the patient's tumor obtained by biopsy. In addition to the tumor cells, the tissue from the patient's tumor contains a mixture of other cell types, including T cells, macrophages, and natural killer cells. The T cell population includes *regulatory T cells* (*Treg*), a $CD4^+$ cell responsible for acting as a brake on the T cell stimulation elicited by its cellular cousin, the $CD4^+$ T helper cell, which orchestrates the immune response.[306]

Macrophages are fascinating cells with the ability, like T cells, to morph into different phenotypic forms under the influence of specific cytokines. The gene expression patterns following cytokine stimulation subsequently determine the functional identity of the local macrophage population. This phenomenon, in which biochemical stimulation of an immune cell determines the cell's phenotype, is called *polarization*. Macrophages can be polarized into *M1* and *M2* populations. M1 macrophages participate in immune surveillance and the destruction of cells bearing non-self-antigens. The difference between macrophages and T cells is that the former is a part of innate immunity, and therefore macrophages do not need to be primed with antigen. They utilize a different recognition mechanism than the cells of adaptive immunity.

Rather than enhance the immune response like their M1-polarized cousins, M2-polarized macrophages engage in tissue remodeling of wounds, where they secrete pro-inflammatory cytokines to recruit the appropriate cells to the wound. They also express elevated levels of matrix metalloproteinases to repair the extracellular matrix. In cancer, this M2 macrophage function slices open the ECM, assisting the tumor cells during their invasion of neighboring tissues.

With a patient biopsy in hand, the tissue is carefully teased apart and subjected to enzymatic treatments that break down the connections of the resident cells to each other and the extracellular matrix. The separated cells are placed in cell growth media in tissue culture flasks. Upon addition of interleukin-2, the T cells expand rapidly, overgrowing the other cell types in the sample (which are not stimulated by IL-2), including immunosuppressive cells (such as Treg cells and M2-polarized macrophages) and even the cancer cells themselves. The final culture contains cytotoxic CD8$^+$ and helper CD4$^+$ T cells with little or no contamination from other hematopoietic cells. After about 5-7 days, a screening procedure can select T cells with specificity for the tumor cell.[307] The cell samples from the screening are next grown in tissue culture (again in the presence of IL-2) until the cell population expands into the number needed for a therapeutic dose. A T cell population of about a hundred billion (10^{11}) T cells is generated for re-infusion back into the patient.[308]

Work performed in the early 2000s in Rosenberg's lab showed that pretreatment of the patient to deplete the lymphocyte count prior to infusion (*lymphoablation*) can significantly increase the survival time of the IL-2-activated T cells that are reinfused into the patient, thereby enhancing the effectiveness of the treatment about tenfold.[309] The patient's pretreatment before receiving an intravenous infusion of potentially disease-altering engineered T cells is a tortuous process of chemo- and/or radiation therapy that eliminates the resident lymphocytes. In 2011, Rosenberg and his colleagues at the NCI published clinical data showing the achievement of long-term remissions (known as *durable remissions* in medical jargon) in a study of melanoma patients.[310] In the three-armed study involving 93 patients, the clinical outcomes of TIL therapy following lymphoablation were reported.

In the trial, all patients were subjected to a chemotherapy regimen with or without additional radiation (at either a high or a relatively low dosage). The radiation treatment provided effective ablation of lymphoid cells (B, T, and NK cells) while sparing the myeloid cells—including erythrocytes (red blood cells), platelets, neutrophils, and macrophages.

The trial outcomes showed that the *objective response rate (ORR)*—that is, the percentage of patients who experienced tumor shrinkage—was higher in the heavily irradiated group (72%) than in the group receiving a lower radiation dose (52%). The ORR rate was 49% for the patients receiving chemotherapy alone without additional radiation. These were highly encouraging results for a disease which, at the time of the trial, had a five-year survival rate of 5%. Elimination of the tumor—known medically as *complete regression (CR)*—was seen in only 2-3% of the patients.

In this trial, twenty patients experienced a complete regression, where there is no evidence of tumor cells at the original tumor site(s) following treatment. Nineteen of those patients were still free of measurable disease three years after treatment. While heavier lymphoablative treatment appeared to impact objective responses, there was no correlation between the various pretreatment regimens and the likelihood of a complete regression. The results in the responders were impressive: when three- and five-year survival probabilities were calculated, the patients who experienced complete regression (also called a *complete response*) had 100% and 93% three- and five-year survivals, respectively, versus 36% and 29% for the entire patient group, including non-responders.

Most amazingly, the persistence of the T cells in the patients improved remarkably from earlier studies done in the absence of lymphoablation. Whereas a vanishingly small 0.01% of the transferred T cells were alive seven days post-infusion in a 1988 melanoma study,[311] 75% of the transferred T cells persisted in the circulation 6 to 12 months post-infusion in some patients in the 2011 trial. Such persistence was a remarkable achievement, not to mention essential for therapeutic efficacy.

TIL therapy remains an active field of clinical investigation for treating metastatic melanoma more than a decade after Rosenberg demonstrated its potential in the clinic. The attraction of the TIL method is that by sourcing the T cell population directly from the tumor tissue, there is likely to be a population of tumor-specific T cells in the TME. In addition, as many tumors contain more than one neoantigen—and melanoma is a high mutational-burden cancer—the TIL population likely contains multiple neoantigen-specific clones of T cells.

This feature differentiates TIL from the other forms of ACT (such as TCR- and CAR-T engineered T cells) discussed later in this chapter, where only one) neoantigen is targeted by the therapy. Here "the more, the merrier" certainly applies, as it is far easier for the tumor to escape immune recognition due to mutation of a single target antigen than when the therapy targets multiple tumor antigens.

For melanoma patients who have failed other therapies, TIL continues to offer hope. In a small feasibility study in the Netherlands, in a cohort of 10 patients, half of the cohort demonstrated an objective response, with two complete responses that have been ongoing for seven years.[312] For patients in the terminal phase of a devastating cancer, these are astounding results by any measure.

For reasons not understood at the time, extending the TIL technology's clinical success in treating melanoma to other cancer types was challenging. There was something special about melanoma that facilitated the tumor-killing power of lymphocytes derived from the tumor. It would take the genomic revolution wrought by the Human Genome Project and breathtaking advancements in DNA sequencing technology to explain why melanomas, of all cancers, are so susceptible to immunological destruction using tumor-infiltrating lymphocytes.

Because melanoma is driven by UV damage from sun exposure and UV light is a potent mutagen of DNA, melanomas tend to have large numbers of tumor-specific mutations in their genomes. The high mutational burden leads to an array of tumor-specific T cells, each variety uniquely qualified to bind to a specific neoantigen in the tumor.

As seems logical, multiple kinds of tumor-specific T cells are likely to neutralize a growing tumor far better than T cells derived from a tumor with a small set of mutations and, therefore, a limited variety of tumor-specific lymphocytes. Thus, while promising results have been achieved for TIL therapy in a high-mutational load tumor, such as lung cancer (specifically, a type called non-small cell lung cancer, found in the majority of smokers diagnosed with lung cancer), clinical outcomes for low-mutational- load cancers like pancreatic cancer have thus far been disappointing.

Difficulties in generating consistently effective TILs for other forms of cancer prompted work on alternative forms of ACT for researchers in the field. At NCI, Rosenberg's lab continued with the TIL work while expanding their ACT work to another form of engineered T cell. The core idea of this research was that the specific interaction between an antigen and the T cell receptor can be exploited by engineering a TCR specific for a tumor antigen into the T cells in the immune infiltrate of the patient's tumor.

The caveat here is that the recognition of the antigen by T cells depends on the association of the antigen with a specific HLA protein on the antigen-presenting cell, a phenomenon called *HLA restriction*. Because of the inherent differences in HLA proteins between individuals, preparing an engineered TCR is far more complex than preparing TILs. While the latter requires only the isolation and expansion of patient T cells, developing a TCR-based therapeutic requires a design that presents the neoantigen to the TCR in the context of the patient's HLA protein profile.

To achieve this, the sequencing data obtained from a tumor biopsy is fed into data processing software that uses molecular modeling and predictive algorithms to evaluate the binding strength (biochemists call it *avidity*) of human antigens to the patient's HLA proteins.[313] The software can compare the results from the patient's tumor antigens to those available for the patient's HLA profile in open access databases. This analysis can identify changes in binding responses indicative of a potential neoantigen that can serve as the basis for the design of a patient-specific (and HLA-restricted) engineered TCR.

It sounds like a needle in a haystack wrapped in a black hole, but herein lies the enormous power of high-sensitivity genomic sequencing (known as *deep sequencing*) coupled with computer algorithms that can detect differences at the level of one amongst billions.

The genes that code for the antigen-binding domains of the TCR from tumor-specific T cells can be identified and copied using the Polymerase Chain Reaction. These genes are incorporated into the patient's T cells in a process known as *gene transfection.* In this process, circular pieces of DNA called *plasmids,* which contain the genes for the antigen-binding domains of the neoantigen-specific TCR, are introduced into the cells using chemical techniques or electroporation. The latter method uses an electric current to transiently create pores in the cell membrane through which the plasmids can slip.

The T cells containing the TCR gene plasmids can be grown in culture (often in the presence of IL-2) to expand the population. When sufficient cell numbers are obtained, the engineered T cells can be re-infused into the patient from which they came. The ability to make clinical doses of these special T cells, so cleverly engineered to express the tumor-specific TCRs on their surfaces, was far beyond our wildest dreams in the pre-genomic days.[314] The fact this has all happened in the span of a single generation (and a single career in the biopharmaceutical industry) is a testament to the startling speed of technological change in the biological sciences in the past half-century.

Important lessons were learned about the use of TCR-engineered T cells in clinical trials. There was unmistakable evidence of clinical improvement in some patients, particularly melanoma patients. Even then, the anti-tumor effects were short-lived, and the treatment was not (in most cases) sufficiently efficacious over time. There were potential weaknesses in the approach that would need to be addressed.

Most significantly, the inherent genetic instability of tumor cells vastly enhances their mutation rate and therefore alters the immunological landscape over time. There is always the potential for loss of the neo-antigenic signal that once drove the immune response. Antigen loss from the target can result from a mutation that renders the antigen invisible to adaptive immunity. In some cases, there is a complete loss of the targeted antigen due to the downregulation of the antigenic protein as the cancer cells adapt to reduce their visibility to immunosurveillance.

In some cases, tumor cells can down-regulate their expression of MHC genes to hide from the immune system, and this can circumvent recognition of the targeted neoantigen in the absence of the HLA proteins needed to present antigens to T cells. In addition, tumor cells have a nasty habit of downgrading the efficiency of antigen presentation in other dastardly ways.

There had to be a better approach to creating effective T cell-based cancer therapeutics.

※※※※※

Following his graduation from the U.S. Naval Academy in 1975, Carl June attended the Baylor College of Medicine and graduated with the intention of pursuing a career as a practicing physician in internal medicine and oncology.

In medical school, June worked in the rapidly developing field of immunology, with a focus on autoimmunity. He developed this interest, at least in part, because his mother suffered from lupus erythematosus. Lupus is an insidious and poorly understood autoimmune disease that is difficult to diagnose definitively and even harder to treat successfully. Existing treatments rely on systemic immunosuppression with steroids and other anti-inflammatory agents, which can help dampen the progression of autoimmune disorders. There is, however, a significant downside: the systemic suppression of immunity increases the risk of both infection and cancer.

After medical school, June's interest in immunology and cancer led him to a fellowship in transplantation biology at the Fred Hutchison Cancer Center in Seattle (1983-1986). The newly graduated Dr. June found himself in the exciting new field of bone marrow transplantation, which was under active clinical development for the treatment of leukemia and other blood disorders.[315]

During his fellowship, the formidable power of the immune system was illustrated for June by the tragic loss of a patient to graft-versus-host disease following a bone marrow transplant. Upon recognizing the transplant as foreign, the patient's immune system launched a blistering attack against it. The devastating liver damage that followed a grueling three-week course of treatment was a grim lesson for Dr. June and the medical community. Even after administering potent immunosuppressive drugs, June noted, "That taught me, in literally just a few weeks, that the immune system is capable of destroying."[316]

After completing his fellowship, June founded the Immune Cell Biology program at the Naval Medical Research Center, where his group worked on engineering patient T cells for resistance to infection by the human immunodeficiency virus, the causative agent of AIDS.

These experiences, in which Carl June witnessed the power of the immune system to both heal and destroy, altered his career path from clinical physician to medical researcher: "I got the idea that I could maybe make a larger contribution by working in a lab rather than seeing patients one by one,"[317] June noted when he left clinical practice in 1999 for a position at the University of Pennsylvania. He is still at Penn today, where he is Professor of Immunotherapy and Director of both the Center for Cellular Immunotherapies and the Parker Institute for Cancer Immunotherapy at the University's Perelman School of Medicine.

Though they showed promise in two types of cancer, the TCR technology suffered from the need to engineer the antigen recognition process in the context of specific human HLA proteins. This is an arduous procedure that is technically demanding and time-consuming. Dr. June capitalized on his experience in T cell engineering for both HIV and cancer to vigorously pursue an emerging form of cell therapy that circumvents the weaknesses of engineered TCRs as therapeutic agents.

June's lab at Penn sought to build upon the successes of engineered TCRs. Realizing the process could be simplified by avoiding the requirement for HLA restriction in the design of T cell therapeutics, they turned their attention to a new type of engineered T cell that was starting to attract interest in the cancer research community.

Design of a therapeutic T cell that engages with its target independent of HLA restriction required a completely different approach than that used in the development of TCR-engineered T cells. TCR-T cells interact with tumor antigens during antigen presentation; they do not directly engage with the cancer cell. Rather, they rely on the ability of the TCR-T cell to respond to stimulation by tumor antigens.

To act on the tumor in the absence of HLA restriction, an engineered T cell would need to engage with the cancer cell. Without engaging the TCR there was no other conceivable way to target the cancer cell

Specific design considerations apply. To engage with the cancer cell, the engineered T cell needs to bind to a protein on its surface. In addition, the engineered T cell needs to not only bind the cancer cell, but it also must elicit a biological signals upon antigen binding that activate T cell killing and stimulates T cell proliferation (as in the case of TCR stimulation upon antigen binding). By engineering these capabilities into T cells targeted at a tumor antigen, it might mimic the functions of the T cell receptor without the need for interaction with the patient's HLA antigens, avoiding the time and expense required to create an HLA-specific TCR.

The idea for designing an engineered T cell that can recognize the antigen that primed it (known to immunologists as its *cognate antigen*) even in the absence of HLA-dependent TCR binding went back over three decades. Shortly after the publication of the T cell receptor's structure in the mid-eighties, Israeli immunologist Zelig Eshhar and his research team at the Weizmann Institute of Science envisioned a way to provide T cell antigen specificity independent of the HLA system.

Utilizing the monoclonal antibody preparation method pioneered by Kohler and Milstein, the Israeli research team created an antibody specific for a target on tumor cells. Since only the antibody-binding function was required—and not the effector functions that reside in the Fc region of the antibody (the stem of the Y-shaped antibody, Figure 2)—the engineered receptor need only contain the antigen-binding domains of the antibody:

the variable regions of the light and heavy chains (V_H and V_L domains, respectively).[318]

In cell culture experiments, Eshhar and his team engineered a T cell receptor in which the (antibody-like) antigen-binding domains of the receptor were replaced with the variable regions of a monoclonal antibody. The results showed that the T cells in culture were able to properly assemble the variable regions at the cell membrane in place of the TCR antigen-binding domains. In addition, the engineered receptor could recognize its antigen and, upon binding, stimulate T cell activation.

When the T cell receptor binds its cognate antigen, the receptor complex and other proteins on the T cell surface interact with other proteins on the antigen-presenting cell (APC) surface (Figure 14).[319] In this process, the biochemical interactions between the participating proteins stimulates the T cell to proliferate while dampening the activity of proteins on the T cell surface that act as "brakes" on the T cell to regulate and limit the T cell's destructive power.

Taming Cancer

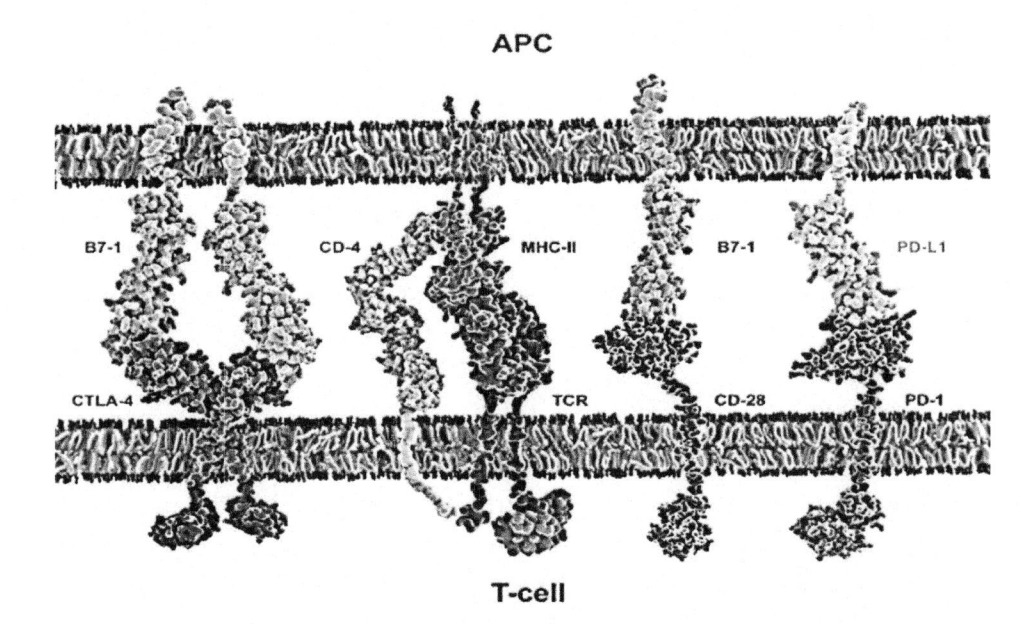

Figure 14. Co-stimulatory interactions of T Cells

Figure 14 shows the protein interactions between a CD4⁺ (helper) T cell and an APC. CD4 is associated with the T cell receptor complex, where it plays a role in mediating the interaction between the T cell and the APC. Upon antigen binding, a protein called *B7-1* on the surface of the APC cell binds to a protein on surface of the T cell called *CD-28*, triggering T cell activation. CD28 is one of several known *co-stimulatory proteins* on the T cell surface (others not shown in the figure). Without co-stimulation, the T cell cannot be activated.

Other proteins on the T cell surface function as brakes on the immune system to prevent the over-stimulation of T cells. Over-stimulation can result in the destruction of normal cells and, potentially, in autoimmunity. Two of these proteins, called CTLA-4 and PD-1, have received significant attention from the cancer research community. These proteins and the cancer therapeutics developed to leverage their biological functions to treat cancer will be discussed in Chapter 14.

Living Drugs

Over the past two decades, multiple laboratories, including both the Rosenberg lab at NCI and Carl June's team at the University of Pennsylvania, have developed and evaluated in patients engineered T cells with both the antigen-binding and co-stimulation properties of the T cell receptor. This type of engineered T cell is known as a *Chimeric Antigen Receptor T cell* (*CAR-T*). The structural design of the CAR-T technology is shown schematically in Figure 15.

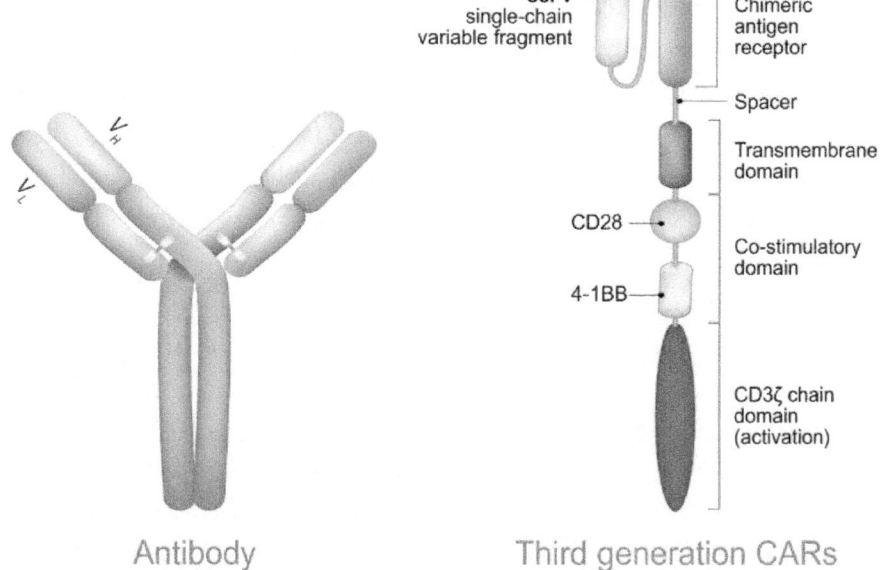

Figure 15. An antibody and a CAR-T construct

Chimeric antigen receptor T cell

scFv single-chain variable fragment — Chimeric antigen receptor

Spacer

Transmembrane domain

CD28 — Co-stimulatory domain

4-1BB

CD3ζ chain domain (activation)

Antibody — Third generation CARs

The truncated form of an antibody containing only the V_H and V_L domains is called a *single-chain variable fragment (scFv)*. Unlike the case of the intact antibody, which is comprised of two copies each of two distinct types of protein chains (heavy and light chains), the scFv is a single protein containing the variable regions of one heavy and one light chain linked together by a flexible amino acid sequence. This sequence, which biochemists call a *linker sequence*, provides the flexibility needed for the association of the V_H and V_L domains, which together form the antigen-binding site. This binding site, which is outside of the cell (extracellular), is connected to a transmembrane domain that snakes through the cell membrane and attaches to one or more co-stimulatory domain(s) inside the cell. The transmembrane domain undergoes a structural change upon antigen binding at the cell surface that triggers the co-stimulation event inside the cell.

Presently, three generations of CAR-T constructs are being assessed in patients. The first CAR-T constructs, which contained only a TCR-associated membrane protein called the *CD3 zeta chain* for T cell activation (first generation CAR-T), did not produce significant clinical results. Further iterations included the addition of one (second generation) or two (third generation) co-stimulatory domains in addition to the T cell receptor's zeta chain. Amongst the multiple co-stimulatory proteins used by T cells, two co-stimulatory proteins, CD28 and a protein called 4-IBB, have been the major focus of clinical research.

CAR-T technology is a fusion of the specificity of antibodies (produced by B cells) and the cytotoxic properties of T cells—a chimera of the two arms of the adaptive immune system. Chimera is derived from the Greek word for a mythical monster comprised of multiple animal parts. Unlike the mythical monster, chimeric antigen receptor technology is real and, to the great fortune of the multitudes suffering from metastatic cancer, this advancement is already making a difference in the lives of cancer patients.

The first published therapeutic success of CAR-T technology for the treatment of leukemia was achieved by Steven Rosenberg and his team in 2010. In 2011, Carl June and his clinical partner, Penn oncologist David Porter, published the results of CAR-T treatment of three patients with chronic lymphocytic leukemia (CLL) who were out of treatment options. Two of the patients had a complete response, with no evidence of the disease in their bodies. This was exciting stuff!

The cancer researchers secured a large grant from the National Cancer Institute to continue their work, resulting in a partnership with Novartis to commercialize the new cancer therapy.[320] In 2017, two CAR-T therapies (Kymriah® from Novartis, and Yescarta® from Kite Pharma) were approved by the FDA for the treatment of B cell malignancies (leukemias and lymphomas) in both adult and pediatric patients. Approvals in the EU, UK, and Canada followed the next year.[321] Additional CAR-T products were subsequently approved for clinical use for B cell malignancies. As of April 2024, there are six CAR-T therapeutics on the market in the United States, and hundreds more are under development.

Even under the best of circumstances, CAR-T therapy can be a brutal experience for the patient. Pretreatment with highly toxic chemical agents to dampen or destroy resident lymphocytes (thereby clearing the decks for the therapeutic cells) is dangerous, resulting in a highly immunocompromised patient. Infusing engineered T cells can lead to severe and even fatal consequences due to the risk of overstimulation of the immune system.

Not all patients are able to risk this ordeal. Herein lies the best rationale for having multiple therapeutic approaches for devastating illnesses, as the course of therapy for these diseases depends on the patient's particular circumstances. This is why personalized treatment with multiple options tailored to the individual patient's needs remains a major goal of cancer drug development.

The most common of the dangerous side effects of CAR-T treatment is a condition called *Cytokine Release Syndrome* (*CRS*, also known as a *cytokine storm*). In cases of CRS, over-stimulation of the immune system causes the release of high levels of multiple cytokines that can make the patient very sick and, if not controlled, CRS can lead to death.[322] Today, treatment of CRS with an antibody that blocks the activity of the immunostimulatory action of the cytokine interleukin-6 (IL-6) is an effective means of treatment of CRS in some patients.

Carl June first suggested this course of treatment to obviate a potentially deadly cytokine release during the treatment of the first pediatric leukemia patient in 2012, a six-year-old girl named Emily Whitehead. Emily was suffering from end-stage leukemia, and her prospects were bleak. Her leukemia had resisted multiple therapies, but she responded to the CAR-T treatment after a harrowing battle with CRS that nearly killed her. Now, more than a decade later, Emily is a healthy teenager.[323]

Significant clinical challenges remain with CAR-T therapy. In addition to CRS and other serious side effects (including damage to the heart, lungs, liver, and kidneys, as well as seizures). The lack of persistence of the engineered T cells in the patient has been a barrier to treatment in many cases. Patient relapses occur when the therapeutic target, a B cell surface protein called *CD19*, is downregulated (or mutated) by cancer cells. In an international trial of CAR-T for the treatment of acute leukemia, relapse due to the loss of CD19 binding occurred in 28% of treated patients.[324] As it turns out, CAR-T therapy is also vulnerable to loss of the target, as even the highly expressed CD19 protein that is critical to signaling in B cells can be jettisoned by cancer cells as they adapt to therapeutic intervention.

The application of the CAR-T approach to treat solid tumors has been challenging due to multiple factors that increase the difficulty of treating them relative to blood cancers. For the treatment of B cell malignancies, the engineered cells are readily delivered into the circulation by intravenous infusion. For the treatment of solid tumors, delivery of the cells to the tumor bed presents enormous challenges. "Compared to hematological malignancies, solid tumor CAR-T-cell therapy is limited by the ability of CAR-T cells to traffic to and infiltrate solid tumors as the immunosuppressive tumor microenvironment and physical tumor barriers such as the tumor stroma limit the penetration and mobility of CAR-T cells."[325]

A significant barrier to success for CAR-T treatment of solid tumors rests on the difficulty of finding an antigen on the tumor cells that is tumor-cell specific, as many of the tumor antigens on solid tumors are often expressed on normal cells. This is also true with CD19-targeted CAR-T therapy for hematological disorders, as CD19 is also found on normal B cells.

As a result, the B cell population is obliterated by treating with anti-CD19 CAR-T cells. The immunoglobulin (IgG) population can be replaced in patients using intravenous administration of human IgG that is obtained from healthy donors. Over time the B cell population is replaced by hematopoiesis as the bone marrow recovers from the treatment.[326]

In a solid tumor, an antigen target shared with normal cells is likely to cause normal cell destruction, a phenomenon called an "on-target, off-tumor effect." While the therapeutic is locating the correct target, that target is on cells unrelated to the genesis of the disease. These effects can cause significant toxicity in patients, and the search for tumor-specific antigens continues for many tumor types.

Today there are hundreds of clinical trials using the CAR-T approach for both hematological malignancies and solid tumors. Unlike other forms of human therapeutics, CAR-T cells expand in the patient during treatment, creating what is hopefully an effective and persistent living drug.

Through their landmark achievements in cell-based therapies, Steven Rosenberg and Carl June ignited a revolution in cancer therapy that is only in its early days. These two visionary scientists showed that by engineering T cells to harness their destructive power, it is possible to transform that power into a potent and potentially lifesaving weapon against the unruly rogue cells that drive human cancer.[327]

Chapter 14

The 21st Century Magic Bullets

In 1996, a short paper appeared from the laboratory of James Allison.[328] A native of Alice, Texas, a small city in the southeastern part of the state 45 miles west of Corpus Christi, his scientific interests were focused on a question that has baffled biologists for decades: how do cancer cells evade the immune response during their progression from a primary tumor to advanced metastatic disease?

Immunologists knew at the time that binding of the T cell receptor to an antigen during antigen presentation is a necessary, though not sufficient, condition for activation of the T cell. Allison's lab at the University of California-Berkeley reported in *Science* that a protein called *cytotoxic T-lymphocyte-associated protein 4 (CTLA-4)* inhibits the activation of T cells. The experiments leading to the discovery of the role of CTLA-4 arose from work in Allison's lab on T cell co-stimulation.

Previous work had shown the presence of key mechanisms that enhance the T cell response. There appeared to be multiple co-stimulatory mechanisms at play, creating a complex signaling network responsible for the regulation of T cell activation. Allison's group noted in the 1996 paper's introduction, "The most important of these co-stimulatory signals appears to be provided by the interaction of CD28 on T cells with its primary ligands B7-1 (CD80) and B7-2 (CD86) on the surface of antigen-presenting cells."[329]

Allison's key insight was stated thereafter: "Therefore, even though most tissue-derived tumors may present antigen in the context of MHC molecules, they may fail to elicit effective immunity because of a lack of co-stimulatory ability."[330] The requirement for co-stimulation led Allison to the realization that "one reason why it was very difficult to actively mobilize T cells to attack cancer cells" is that "it isn't enough to just try to push them into going, but you have to learn how to suspend the brakes, if you will, at least temporarily, to really realize the full potential of the immune response."[331] Allison developed a hypothesis that the tumor's ability to evade immune detection and destruction by host immunity may rest (at least in part) on a tumor-mediated mechanism blocking the interactions of proteins such as CD28 and B7-1 with T cells, thereby disrupting the co-stimulatory signaling required to fully engage the T cell-mediated immune response.

The Allison lab zeroed in on *cytotoxic T-lymphocyte- associated protein 4 (CTLA-4)*. Like CD28, CTLA-4 on T cells binds to B7-1 on APCs, but with higher affinity than CD28, such that CTLA-4 can push CD28 off B7-1 and take its place. Since CD28 and CTLA-4 compete for binding to antigen-presenting cells, and CD28 provides a stimulatory signal, CTLA-4 might serve the opposite function by blocking the T cell stimulation. In the absence of a "stop signal," the immune response would attack healthy cells, eliciting autoimmunity and tissue destruction.

If Allison had it right, injecting antibodies to CTLA-4 into tumor-bearing mice would prevent CTLA-4 binding to B7-1, thereby allowing CD28 binding to favor T cell activation. By removing the brake on T cell activation provided by CTLA-4 binding, he realized it might be possible to turn quiescent T cells back into tumor killers.

The results in animal studies were dramatic. By blocking CTLA-4 with antibodies, a significant increase in survival and reduction in tumor size was observed in the treated animals for both newly injected and established tumors, with no response evident in the untreated animals.[332]

The human clinical trials demonstrated the utility of using an antibody to block CTLA-4 in several types of cancer. As the binding of CTLA-4 provides a checkpoint on immune activation, Allison's discoveries gave birth to an entirely new field of cancer treatment called *immune checkpoint inhibition*. Monoclonal antibodies directed at immune checkpoints are called *immune checkpoint inhibitors*. By providing the means for releasing the brakes on T cell stimulation, this discovery revolutionized the treatment of several types of solid tumors, including lung cancer and melanoma.

On the other side of the world from Allison's Texas lab, Tasuku Honjo had, since the early 1990s, been actively investigating a different immune checkpoint protein called *programmed death-1* (*PD-1*), which Honjo had discovered in populations of dying T cells. The genetic expression of the PD-1 protein increased in T cells undergoing apoptosis (aka programmed cell death), a phenomenon that provides the origin of the protein's name.

In a 1992 paper, Honjo's lab reported that activated T cells express elevated levels of PD-1 on their surfaces. Biochemical characterization studies revealed that PD-1 is a transmembrane protein with a genetic fingerprint showing it is a member of the immunoglobulin family of proteins. Further studies showed that PD-1 expression is also enhanced in activated B cells.

Structural studies demonstrated similarities between PD-1 and CTLA-4. Like PD-1, CTLA-4 is a transmembrane protein with properties that place it in the immunoglobulin family. The portion of CTLA-4/PD-1 protruding from the membrane resembles the variable regions of an immunoglobulin. The structural similarity between CTLA-4 and PD-1 could not be a coincidence, and likely reflected a similarity in function.

The next groundbreaking discovery was revealed in a paper in the *Journal of Experimental Medicine* in 2000. Honjo's team showed that, like CTLA-4, the binding partners of PD-1 on APCs are also members of the B7 gene family. A paper published the previous year by Honjo's lab had shown that disrupting the expression of PD-1 in mice resulted in an autoimmune disease syndrome reminiscent of human lupus, with inflammatory damage to multiple organs. This suggested that PD-1 had a similar function as CTLA-4, serving as another brake on the immune response.

Honjo called the B7-related binding partner of PD-1 the *ligand of PD-*

1 (PD-L1). In a clever series of experiments, the 2000 paper laid out the data showing that when PD-1 binds to PD-L1, T cell activation is dampened. This finding confirmed that the PD-1/PD-L1 pathway serves as a checkpoint on the immune response.[333]

Checkpoint inhibitor antibodies are currently licensed for about a dozen types of cancer. Never has a single class of human antibodies shown so much potential for slowing down and even reversing some cancers. While the current generation of checkpoint inhibitors achieves long-term remissions in only about 20% of treated patients, the results can be dramatic. Former President Jimmy Carter's melanoma had metastasized to his liver and brain when he began treatment with radiation and the checkpoint inhibitor pembrolizumab (Keytruda) in 2015. Months later, he was cancer-free. President Carter, our longest-living ex-President at 99 years of age, is still free of the advanced cancer that threatened his survival.

An exciting new therapeutic modality for cancer has emerged that leverages a half-century of groundbreaking discoveries in virology and immunology. *Oncolytic viruses* are genetically engineered viruses that can infect, replicate, and destroy cancer cells. By engineering the viral genome to control the virus's biological behavior, it is possible to target the virus's destructive power only at cancer cells, leaving normal cells intact. It sounds ambitious—even a tad like it comes from the world of science fiction—but oncolytic viruses have been an active area of cancer research for decades.

While this therapeutic approach shows great promise, there is currently only one oncolytic cancer drug licensed in the U.S., E.U., and Australia. This oncolytic viral therapy for treating advanced melanoma, developed at University College London and later commercialized by a small British biotechnology company called BioVex, carries a tongue-twister of a name: *talimogene laherparepvec* (also known as *TVEC*). Approved by the FDA for the treatment of advanced melanoma in October of 2015 under the trade name *Imlygic®*, the drug was later approved in the European Union and Australia.

TVEC is an engineered form of the herpes virus *Herpes Simplex-1 (HSV-1)*. This virus causes painful blisters on epithelial cells on body surfaces during active infection. It can also lie dormant in nerves, with intermittent outbreaks that tend to occur when the infected individual is under stress. While HSV-1 is commonly found on oral surfaces, it can also cause genital herpes, which is most often due to infection with the closely related *Herpes Simplex Virus-2 (HSV-2)*.

To render the engineered virus safe and effective for treating cancer, the viral genome was engineered to permit replication in cancer cells while preventing replication in healthy cells. This neat bit of genetic sleight of hand can is enabled by the removal of a viral gene that overrides a normal host cell defense against viral infection.

At the first signs of infection of normal cells, both the infected cells and subsets of immune cells increase the expression of interferons. The presence of interferons triggers a coordinated immune response against the invader, with further secretion of pro-inflammatory cytokines that activate the immune system. During immune activation, subsets of immune cells counterbalance immune activation by secreting anti-inflammatory cytokines. This complex interaction maintains the delicate balance between immune activation and immune over-stimulation. Over-stimulation can lead to a breakdown of immune communication with concomitant uncontrolled release of cytokines in a potentially lethal cytokine storm.[334]

As part of the host anti-viral response, healthy cells produce a protein called *Protein Kinase R*, or *PKR*, a signaling molecule involved in activating pathways that inhibit cellular proliferation and viral replication by shutting down host protein synthesis. Some viruses have the capability of nullifying the PKR-mediated anti-viral response. Herpes virus is one of these.

The herpes genome contains a gene called *$\gamma 34.5$* (*gamma-34.5*) that codes for a protein that inhibits the PKR pathway and thereby blocks the host cell's normal ability to shut down protein synthesis following infection. This gene is deleted in the TVEC virus, so the virus cannot shut down the host's anti-viral response. As a result, normal cells can exert a potent anti-viral response that prevents active infection by interfering with the virus's ability to make viral proteins. Cancer cells, on the other hand, provide a fertile breeding ground for the virus, as they are highly deficient in many normal host cell processes, including resistance to viral infection. This key biochemical change to the engineered virus renders it highly selective for the infection of cancer cells.[335]

In addition to the deletion of $\gamma 34.5$, TVEC has two other changes in the viral genome. First, there is the deletion of a gene that codes for a viral protein, which reduces the ability of the infected cell to form antigen-MHC class I complexes. The deletion of this gene prevents the virus from diminishing the presentation of tumor antigens to T cells, knocking out the virus's natural ability to interfere with antigen presentation.

The second change to the viral genome involves the insertion of a gene for a human cytokine called *granulocyte macrophage colony-stimulating factor* (*GM-CSF*). This protein is an immune modulator that, amongst other functions, recruits antigen-presenting cells to the site of its expression—in this case, virally-infected tumor cells. GM-CSF signaling provides a kick-start that activates the innate and adaptive immune responses to the virus by stimulating the proliferation of neutrophils, macrophages, and dendritic cells.

TVEC is delivered to patients using *intralesional injection*—a fancy way of saying that for treating melanoma, the virus is injected directly into melanoma lesions on the skin (or scalp). The injections, given as a series over time, lead to massive viral infection in the melanoma cells. Most of these cells die by about a week post-injection. Due to the removal of the viral gene $\gamma 34.5$, normal cells can resist viral infection, and tissue destruction goes right up to the boundaries of the lesions and stops there. Remarkable.

A dead cell is a leaky cell; the cell membrane bursts, and the cellular contents are disgorged into the extracellular space. There, immune cells can "pick up the scent" of the aberrant cells (and cellular constituents) they had previously been unable to detect. This mixture of viral and tumor cell proteins contains neoantigens from driver mutations.

When dendritic cells present these neoantigens to T cells, there is the potential to establish an immune response that begins in the tumor lesions and subsequently reaches deep inside the body to destroy metastatic lesions in the organs. In this manner, the oncolytic virus approach meets the most fundamental requirement of successful immunotherapy: to turn what is known as a "cold tumor"—one in which the cancer is invisible to the immune system—into a "hot tumor" that elicits an activated tumor-specific immune response.

The lysis of the tumor cells by oncolytic viruses releases the entire *proteome*—the entire protein content of the cell, including the proteins from inside the nucleus, those outside the nucleus in the cytoplasm, the proteins that comprise the cytoplasmic structures (organelles), and the proteins that span the cell membrane into the extracellular space. Because this therapeutic modality exposes the entire proteome to the immune system, with all neoantigens in play (regardless of their location in the cell), the loss of expression of a single neoantigen presents a lower risk of therapeutic failure compared to a therapy targeted with only one or a limited number of neoantigen(s).

Conversely, in CAR-T therapy, where the T cell is targeted at a single antigen on the tumor cell surface, the therapeutic is vulnerable to target loss over time as the cancer cells adapt to the pressure exerted by the activated T cells seeking to destroy them by downregulating the expression of the target. This process is called *immunoediting*.

Because the virus is engineered to cripple its infectivity and pathogenicity (i.e., disease-causing potential) in normal cells, the treatment has a favorable side effect profile compared to many other kinds of cancer treatment. Many patients live with a higher quality of life under oncolytic virus treatment than other late-stage cancer patients. Also, there is no need for the patient to endure a pretreatment that destroys the lymphocyte population, which is required for bone marrow transplants currently used to treat many blood cancers and the various forms of ACT (such as CAR-T) described previously.

Small phase I trials demonstrated the safety of the TVEC virus in several types of cancer, with manageable clinical side effects, including fever, fatigue, and injection site reactions. These side effects are typical for a therapy that induces an immunological response, which includes the production by activated immune cells of various pro-inflammatory cytokines known to induce these effects.

Phase II trials with TVEC were conducted on patients with advanced melanoma, the cancer type that was easiest to treat and showed promise in the phase I trials. In those studies, a state of "stable disease" was achieved in some patients. Stable disease means the tumor burden—the total amount of tumorous tissue—did not increase or decrease significantly in these patients. It was encouraging to see signs of active viral infection in the treated tumors, with little evidence of the virus in the surrounding healthy tissues. Some melanoma patients also showed signs of biologically meaningful levels of GM-CSF expression resulting from the treatment.

In the Phase II clinical program, 50 advanced melanoma patients were dosed with TVEC. Three-quarters of the patients had previously been treated for their disease without success. In this study, about a quarter of the patients (26%) experienced an objective response with a side effect profile similar to that seen in the Phase I studies.[336] Of the 13 responding patients, 8 experienced a complete response, the apparent elimination of the tumor.[337] Most impressively, there was clear evidence of destruction of both primary tumors as well as metastatic lesions, the first unequivocal demonstration of both the oncolytic properties of TVEC and its ability to generate a systemic immune response. Finally, an intriguing, delayed effect was observed: during follow up, two patients who were treated without initial benefit developed complete responses by 24 months post-injection.[338]

In phase III pivotal clinical trials—the phase of clinical study that provides the critical efficacy data on a large group of patients—436 patients with advanced melanoma were split into two groups. The treatment group received TVEC with a higher dosing frequency than was used in Phase II, and the patients in the control group were administered GM-CSF, a licensed drug used for melanoma treatment. Because TVEC itself "creates" GM-CSF as part of the treatment, it is sensible (and required) to compare the impact of the therapy against GM-CSF alone. In this way, the impact of the oncolytic virus (if any) can be differentiated from that of the cytokine it expresses.

In the phase III trial, the treatment's efficacy was defined by what is called the *durable response rate (DRR)*, the percentage of treated patients who develop an objective response that lasts more than six months during the first 12 months of treatment. The results showed a statistically significant 16.3% DRR in the treated group, with a 2.1% DRR in the GM-CSF treatment control group.[339] The overall objective response rates (durable or not) were 26.4% and 5.7% for the TVEC and control groups, respectively, with a similar side effect profile to that seen in earlier trials.[340] There were complete responses in 11% of the TVEC-treated patient population.

The evaluation of the phase III data identified that a subset of patients had a higher response rate than that demonstrated for the entire TVEC-treated population. The trial enrolled patients with various stages of advanced melanoma, from the stage where there is localized spread of the surface tumors and/or spread to one or more nearby lymph nodes (stage III) to the most advanced stage of melanoma, stage IV. In addition to local lymph node involvement, these patients have metastatic lesions at sites distant from the melanoma lesions on the body's surface(s).

The results showed that clinical efficacy was far better for patients in the earlier stages of metastatic melanoma than for later stage patients. Also, patients with higher tumor burdens (who tend to be the latest stage patients) tended not to do as well as those with lower total tumor mass.[341] This meant that if TVEC is given to patients before they reach the end stage of the disease, the efficacy rate might be significantly higher than those seen in the phase III trial.

Current clinical efforts at oncolytic virus development are evaluating a dozen types of viruses targeted at difficult-to-treat solid tumors such as cancers of the liver, lung, pancreas, and prostate, as well as for treatment of deadly brain cancers, including glioblastoma, the most common form of brain cancer.[342] This therapeutic modality, still in its infancy, offers significant promise for the future treatment of solid tumors.

Taming Cancer

The vaccines we are familiar with are designed to prepare the immune system for a future encounter with an infectious agent. These vaccines have become a part of our everyday lives. I recall, as a first grader, proudly lining up in the cafeteria of Ogden Elementary School to receive the new oral polio vaccine invented by Dr. Albert Sabin.[343] Our parents were thrilled that we were receiving protection from the ravages of poliomyelitis, a potentially crippling and even fatal viral pathogen.[344]

Vaccines administered for the prevention of disease are called *prophylactic vaccines*. The list is familiar: polio, measles, mumps, rubella, diphtheria, pertussis (whooping cough), HPV (the cause of most cervical and anal cancers), hepatitis B, influenza, and, most recently, COVID-19. These vaccines are designed to prime the immune system to create immune memory in the B and T cell populations in the event of a future infection.

Prophylactic vaccines contain either inactivated whole viruses or a viral protein (or proteins). The recent introduction of messenger RNA (mRNA) vaccines adds an important new modality to the vaccine development toolbox. As these vaccines can be developed and manufactured significantly faster than protein-based vaccine technologies, they provide a significant advantage in the face of emerging infectious diseases like COVID-19.

It is safe to say that no other medical development in history has saved more lives than these products of more than two hundred years of medical advancement. Without these vaccinations, smallpox—the most pernicious viral disease in history—would still be a large-scale global killer. Our experience with prophylactic vaccination has proven that these medical wonders are essential to the control and sometimes total eradication of dangerous infectious diseases.

Although it is not part of our everyday experience, vaccinations can be used for another purpose besides preventing infectious diseases: they can be administered for treating active diseases. Recall that Coley's Toxin was administered to patients to stimulate the immune system. Like prophylactic vaccines, Coley's Toxin was comprised of biological agents (partially inactivated bacteria) designed to treat active cancer. Thus, Coley's Toxin was a *therapeutic vaccine*, a vaccination for the treatment of active disease.

A therapeutic cancer vaccine is designed not to prime the immune system to prepare for future exposure; rather, the purpose of such a vaccine is to treat an existing cancer. Its role is to awaken the immune system to a tumor's presence and stimulate a potent immunological response against it. To achieve this, the therapeutic vaccine must sufficiently stimulate the immune system's innate and adaptive arms to generate a response that can destroy the tumor over time.

This approach is fraught with difficulties. Most significantly, these vaccines are given in the context of an advanced cancer patient with a compromised immune response. In the TME, regulatory T cells, M2-polarized macrophages, and myeloid-derived stromal cells (MDSCs) secrete immunosuppressive cytokines that diminish T cell function, including *interleukin-10* (*IL-10*) and *transforming growth factor beta* (*TGF-β*). In addition, like other forms of immunotherapy, cancer vaccine efficacy is hampered by immunoediting of the target neoantigen by tumor cells.

Cancer therapeutic vaccine development has taken multiple forms: whole tumor cells, nucleic acid vaccines, and vaccines comprised of protein neoantigens or *tumor-associated antigens*, such as antigens that are usually expressed only during fetal development. These antigens are less preferable because they are often expressed at low levels in normal cells such that targeting them entails a risk of side effects due to on-target, off-tumor effects.

Cell-based vaccines have also been developed using dendritic cells, which are collected from the blood and loaded with tumor neoantigens. Upon reinfusion into the patient, these neoantigen-loaded dendritic cells present their cargo to the immune system to stimulate a response.

Amongst the modalities of cancer vaccines under study, the most promising are DNA vaccines.[345] These vaccines contain the DNA that codes for one or more of the protein antigens on the tumor, preferably tumor-specific neoantigens that have been identified by studying tumor biopsies. The expression of the DNA in the vaccine creates many copies of the tumor neoantigen in the hope of eliciting a robust anti-tumor immune response.

DNA vaccines are under investigation in a broad array of cancers. While promising results have been reported in animals, human clinical trials have yet to demonstrate a vigorous response.[346] Nevertheless, cancer vaccines (including those based on the new mRNA vaccine technology) will almost certainly have a place in cancer treatment. Recent results in human trials demonstrate their potential utility, perhaps not as a stand-alone therapy, but rather one that can work in cooperation with other types of immunotherapies to obtain potent anti-tumor effects.[347]

We will explore such combination therapies in the concluding chapter.

The IgG structure contains two identical antigen-binding domains, one on each "arm" of the Y-shaped structure (Figure 2). Since the binding domains are identical, each antibody can bind up to two antigens of the same type.

Now imagine replacing the binding sequences on one arm of the antibody with sequences specific to a different antigen. This format would allow the antibody to bind antigenic sites on two different proteins. If these antigens are on the surfaces of two different cells, the antibody could bind to one antigen on each cell, thereby forming a bridge between the cells.

Antibodies of this type—known as *bifunctional antibodies*—have been successfully developed for the treatment of hematological (blood) cancers. By binding to a killer T cell with one arm of the antibody, and a surface protein on malignant B cells on the other, bifunctional antibodies can bring the T cells into proximity with the B cells. Once the T cell is properly oriented with respect to the B cell, the T cell can exert its cytotoxic activity and kill the cancerous B cell. Once initiated, the cytotoxic process generates a chain reaction as the tumor-specific T cells secrete cytokines that fuel their proliferation. This creates increasing numbers of cytotoxic T cells ready to zero in on their molecular targets.

The first bifunctional antibody licensed for the treatment of cancer was developed by a small biotechnology company based in Munich called MicroMet. After Amgen purchased MicroMet and its bifunctional antibody portfolio, they introduced into clinical trials a bifunctional antibody that binds on one arm the same CD19 protein on B cells as used in CAR-T therapies. The other arm of the bifunctional antibody binds the CD3 protein component of the TCR in T cells. The binding of the bifunctional antibody to these targets builds a bridge between a malignant B cell and a cytotoxic T cell.

With the generic name of *blinatumomab* and the trade name Blincyto®, this molecule has been highly effective in some patients in advanced stages of acute lymphoblastic leukemia. Today, at universities and biopharmaceutical companies, bifunctional antibodies are under development for treating cancer and other grievous illnesses.

Compared to cell-based therapies, the bifunctional antibody approach to cancer treatment has some notable advantages. Because the antibody can directly elicit tumor cell killing, there is no need for pre-ablation of the lymphocyte population with toxic agents. Another significant advantage is that, unlike CAR-T therapeutics—which are custom-made from each patient's T lymphocytes—a bifunctional anti-tumor antibody is an off-the-shelf therapy that does not require weeks of preparation for each patient. Compared to cell-based therapies, this approach is not as costly and is more readily accessible for treating patients. There is little question that bifunctional antibodies will continue to provide significant benefits in the treatment of cancer and other diseases.

※※※※

Traditional chemotherapeutic agents are delivered into the circulation by intravenous infusion or orally and delivered by the bloodstream throughout the body. This mode of therapy (*systemic therapy*) does not discriminate between one organ system and another in how it distributes the drug inside the patient. The delivery of the medicinal payload exclusively to the diseased tissues (rather than systemically) has been a long-sought aspiration in pharmaceutical science. If this can be achieved, it becomes possible to concentrate the therapeutic agent inside damaged cells while sparing normal cells from its toxic effects.

The development and commercialization of monoclonal antibodies demonstrated that proteins on the cell surface can be targeted by exploiting immunological recognition between an antibody and an antigen. Once this was demonstrated, it was possible to envision a way to specifically deliver a toxic agent to a cancer cell by using a monoclonal antibody as a chaperone capable of bringing the toxin directly to its targeted cell.

Such a therapeutic agent—a monoclonal antibody linked to a cytotoxic chemotherapeutic agent—is called an *immunotoxin*. The idea, which has been around for decades, has long been the focus of Dr. Ira Pastan's laboratory at the National Cancer Institute. As a pioneer in the field of immunotoxins, Pastan has spent more than four decades investigating the use of bacterial and plant toxins for targeted cancer therapy. By linking cellular toxins to monoclonal antibodies specific for tumor targets, this approach increases the potency of monoclonal antibodies as cancer therapeutics.

Pastan's lab focused on using a bacterial toxin from *Pseudomonas aeruginosa* called *exotoxin A*. This exotoxin is a significant contributor to the pathogenicity of *Pseudomonas*; it interferes with host cell protein synthesis. The idea of leveraging the toxicity of exotoxin for cancer treatment grew out of work by a postdoctoral researcher in Pastan's lab named Dave Fitzgerald, who was investigating the mechanisms by which toxins enter cells. As a model for his toxin studies, Fitzgerald chose *Pseudomonas* exotoxin A.

In an interview with the Chinese Antibody Society, Dr. Pastan recalled the excitement in the lab generated by the exotoxin work: "While doing these studies we were impressed with how effective *Pseudomonas* toxin was in killing cells. So, we thought it might be an innovative idea to attach it to a monoclonal antibody to target and kill cancer cells. Around that time, there was a lot of interest in using monoclonal antibodies to deliver radioisotopes, cytotoxic drugs, and toxins to kill cancer cells."[348]

When Pastan first proposed using bacterial toxins for cancer treatment, there was vigorous skepticism amongst immunologists, many of whom predicted the approach would not work. This prediction was as accurate as Admiral Leahy's World War II warning that as an expert in explosives, he knew the atomic bomb would never work. So much for predictions. The best scientists know it is sometimes best to ignore such feedback and press on.

In 2012, Pastan and his clinical partner, NCI's Dr. Robert Kreitman, reported on the results of a Phase I trial of an immunotoxin called HA22.[349] This therapeutic agent, designed to target B cell leukemias, is comprised of a monoclonal antibody to a B cell surface protein called *CD22* linked to the *Pseudomonas* exotoxin. Pastan received exhilarating news when his partner called him early in the trial to report the results from the first patient treated with HA22: "The leukemia count has fallen by 50% and it's still going down."[350] By leukemia count, Dr. Kreitman was referring to the blood panel that measures the numbers of distinct types of white blood cells. Incredibly, that report came after the patient's first dose.

Upon trial completion, Pastan and Kreitman found the results were incredible: of the 28 patients in the trial, 13 experienced a complete remission. Years later, Pastan reported, "Eight or ten years later, some of those patients have survived without any detectable cancer."[351] This remarkable achievement must have been sweet satisfaction for Dr. Pastan and his colleagues, a product of perseverance, diligence, and the wisdom to ignore the naysayers in pursuit of something really cool that you think might just work.

In a global clinical trial completed in 2018 that evaluated the efficacy of the HA22 immunotoxin in patients with a difficult-to-treat B cell leukemia called *hairy cell leukemia*, 80% of the patients went into remission from their cancers, with 41% experiencing a complete and durable response. Wow! When it comes to treating advanced cancers, results like these are rare indeed. HA22 and the development team that brought it to the clinic had hit it out of the park with the bases loaded in the bottom of the ninth.

The HA22 immunotoxin, called *moxetumomab pasudotox* (*Moxe* for short), received FDA approval in 2018 and is marketed by MedImmune (now part of Astra Zeneca) under the trade name Lumoxiti®. A drug more than twenty years in the making, Lumoxiti is a highly effective drug for difficult-to-treat leukemias. Today, other immunotoxins are under investigation for leukemias, lymphomas, and solid cancers. These agents offer hope to cancer patients who have failed other therapies.

A new type of highly potent immunotoxin called an *antibody-drug conjugate (ADC)* has emerged in the past decade. Instead of coupling a bacterial or plant toxin to the antibody, an ADC is comprised of a monoclonal antibody specific to a cancer target antigen linked to a highly toxic compound known as the "warhead." These warheads are far too toxic to be administered systemically to patients.

The toxins chosen for this application are two to three orders of magnitude (100-1000 times) more destructive to cells than chemotherapy drugs that are injected systemically. Because these compounds are so toxic, they must be taken into the cell before releasing their toxic payload. These payloads are toxins that either bind to DNA and cause irreversible breaks in the DNA strands or function as potent inhibitors of the *microtubules* that are necessary for cell movement and cell division (by interfering with the action of the spindle fibers that pull apart the chromosomes during mitosis).[352] These compounds would be lethal at even miniscule doses if administered intravenously without attachment to a chaperone to deliver them inside the targeted cells.

After an antibody binds to a receptor protein on the surface of a cell, the antibody-antigen complex undergoes *receptor-mediated endocytosis,* whereby the antibody-antigen complex is enveloped by a piece of membrane that takes the antibody and its bound antigen into the lysosome. The chemical linker between the antibody and the toxin is not cleaved until the antibody-drug conjugate is engulfed by cells and trafficked to lysosomes. Inside the lysosomes, the acidic pH destabilizes the bond between the antibody and the chemical toxin to release the toxin inside the cell.[353] This allows for concentration of the toxin inside cancer cells after the antibody binds its targeted surface protein.

ADCs overcome some of the clinical limitations of immunotoxins containing toxic proteins derived from bacteria or plants, as they do not contain a foreign protein while reducing systemic exposure to the toxin. The presence of the foreign protein in the immunotoxin leads to an immune response after the second (or sometimes third) treatment.[354] In addition, systemic immunotoxins can damage the endothelial cells lining blood vessels, resulting in a complication called *vascular leak syndrome (VLS)*, which can cause severe and dangerous hypotension (low blood pressure). The extent of VLS is toxin- and patient-dependent. In the case of Lumoxiti and other agents that use the *Pseudomonas* exotoxin, the immune reaction is usually manageable with supportive measures such as steroid treatment.

About 100 ADCs are currently in clinical trials, and several members of this class of cancer therapeutics are licensed and used in clinical practice. These agents target both liquid cancers and solid tumors. This technology, still in its infancy, is "set to represent an important contribution to the future of immuno-oncology."[355]

Chapter 15

The Future of Cancer Medicine

The startling advancements in cancer therapy described in the previous three chapters are products of decades of research in biochemistry, immunology, cell biology, and molecular genetics. For the first time in our lengthy battle against cancer, we have built a solid foundation for effectively incorporating genomic and biochemical data into cancer medicine on a patient-specific basis so that effective therapeutics can be designed that zero in on the right targets for each patient throughout their course of treatment.

Seemingly miraculous cases of long-term cancer remissions, and even apparent cures, have sometimes followed treatment with immunotherapeutic agents. Despite these thrilling developments, significant hurdles remain. For example, the clinical experience thus far with checkpoint inhibitors—the only class of immunotherapeutics with sufficient usage to date to supply meaningful statistics—have only rendered such impressive responses in about 15% of patients. It is too soon to assess the efficacy (not to mention the affordability) in clinical practice of oncolytic viruses, cancer vaccines, bifunctional antibodies, ADCs, and engineered T cells.

The future of cancer therapy rests on the answers to some fundamental questions: Why do some patients respond favorably to treatment, whereas others with similar cancers and risk factors succumb to their disease without responding to these powerful new therapeutic agents? How can we increase the clinical efficacy of these exciting new therapies? Can therapeutic success rates be improved via molecular screening of patients to determine who might respond to a given treatment?

We can now go far beyond characterizing cancers by their location(s) in the body, microscopic appearance, and biological behavior. Using the advanced analytical tools described in previous chapters, we can subject each patient's cancer to a thorough molecular examination. By sequencing the DNA and several types of RNAs (the products of transcription, aka the *transcriptome*), and identifying the collection of proteins expressed by the cancer cells (the *proteome*), customized therapeutic strategies can be devised to target the molecular roots of each patient's cancer.

By monitoring changes in the molecular profile during treatment, we can determine how cancer cells respond to therapy in real time. As a result, we can gather critical information on the development of drug resistance at the molecular level. Based on this data, the treatment regimen can be proactively adjusted to meet the ever-changing dynamics of tumor evolution.

We call this molecular approach to treatment *genomic medicine* or *precision medicine;* when applied in treating cancer, we call it *precision oncology*. The realization of the dream of precision oncology rests on two foundational capabilities that have come of age only recently: (1) the ability to characterize in exquisite detail the molecular characteristics of each patient's cancer and (2) the ability to effectively leverage the accumulated knowledge in cancer biology, immunology, biochemistry, molecular biology, and cellular engineering to design powerful new therapeutics to treat a higher proportion of cancer patients.

The precision medicine approach is not new in cancer care. As an example, Philadelphia chromosome-positive chronic myelogenous leukemia has become a far more manageable cancer due to the availability of the targeted therapeutic Gleevec® for patients with Philadelphia Chromosome-positive CML (Chapter 6). In addition, a class of drugs known as *PARP inhibitors*, which leverage the damaged DNA repair mechanisms of tumor cells to cripple DNA replication, has been effectively deployed in treating breast cancers resulting from mutation of the DNA repair enzymes encoded by the genes *BRCA1* and *BRCA2*.

The discovery that some breast cancers have a driver mutation in the HER2 receptor has transformed breast cancer therapy. The use of HER2 as a marker in tumor tissues in laboratory tests for characterizing breast cancers, along with the use of the anti-HER2 monoclonal antibody Herceptin for treating HER2-positive breast cancers, is an example of the precision oncology approach.

The long history of cancer medicine tells us the use of targeted therapies alone cannot eliminate all the tumor cells in most metastatic cancers. These agents are often highly effective at shrinking tumors and inhibiting their growth, at least for a while. Unfortunately, resistant cancer clones generated during and after treatment can sequester themselves in a quiescent state for prolonged periods measured in years or even decades. There they remain, awaiting the right conditions to emerge from their long slumber. When awakened, they emerge fully reenergized to continue the relentless quest for proliferation characteristic of cancer cells.

The ability to activate the human immune system to recognize and destroy tumors represents a profound shift in the cancer treatment paradigm. Instead of going straight at the cancer cells, these therapies are designed to teach the immune system to do what it does best.

Unlike chemotherapy and many targeted agents, immunotherapy impacts both metabolically active and more quiescent cells that have reduced rates of proliferation and metabolism (such as cancer stem cells). Any cells bearing the protein marker targeted by the treatment are vulnerable to immunotherapeutic agents regardless of their metabolic status. As a result, cancer can now be destroyed from within by a therapy with the potential to eliminate most or all the patient's cancer cells. If some cancer cells are left behind, immunological memory may persist to guard against the future proliferation of the survivors.

Given the ever-growing arsenal of cancer therapeutics, the question arises about how they can best be deployed. In a 2019 interview, Dr. June expressed his hope that immunotherapeutic treatments will become the dominant form of cancer therapy in the future: "This is our hope, too, namely that immunotherapy will be at the front end of cancer therapy for patients with all types of cancer, and it will be achieved using many different kinds of engineered cells targeting cancer and the microenvironment of the tumor."[356]

Inherent in this vision is the realization some patients will not be eligible for treatment with immunotherapeutic agents. It might be due to the nature of their cancer, the presence of other diseases limiting their therapeutic options, the inability to withstand lymphodepletion and other pretreatment regimens, a high vulnerability to the side effects, and/or other medical complications. In these cases, there might be a corollary of the vision: for those patients ineligible for treatment with immunotherapeutic approaches, the ability of the cancer to spread beyond the primary site and establish metastatic lesions can be blocked by manipulating the biochemical pathways that drive tumor progression.

Based on our current knowledge, we can imagine therapeutics that disrupt the cancer cells' metabolism and innate plasticity, impeding the evolution of increasingly invasive cancer cell clones and thereby inhibiting the evolution of tumor cells capable of leaving the tumor bed. For example, there is current interest in therapeutics that inhibit the epithelial-to-mesenchymal transition (EMT) that promotes the invasive characteristics of evolving cancer cells.

In addition, there is interest in drugs that reduce the levels of the matrix metalloproteinase enzymes in the tumor bed that enable cancer cell invasion by slicing through the extracellular matrix in the vicinity of the tumor. Finally, our knowledge of the mechanisms of metastasis demonstrates the potential of therapeutic agents that interfere with the transit of disseminated tumor cells in the circulation and/or prevent the creation of "fertile soil" in pre-metastatic niches, thereby "poisoning the soil" and precluding the successful implantation and growth of metastases. By so doing, it might be possible to disrupt the ability of tumors to manipulate normal cells and flip the dynamic by gaining the ability to pharmacologically manipulate tumors.

Using these strategies, it might be possible to turn cancer in these treatment-resistant patients from an acute, life-threatening ailment to a chronic disease that can be managed over prolonged periods like other chronic illnesses, such as diabetes, COPD, and rheumatoid arthritis. Due to the lower toxicity of targeted therapeutics and immune system modulators relative to systemic cytotoxic therapies, patients would experience a higher quality of life than is typical for today's sufferers of advanced cancer.

We distinguish here between cytotoxins that are injected into the bloodstream or given orally (as in most traditional chemotherapies) and the direct delivery of cytotoxins to the tumor cells, as for immunotoxins and ADCs. Reductions in the use of chemotherapies that expose cancer patients systemically to potent cellular toxins would provide an enormous leap forward in the treatment of human cancer.

Cytotoxic drugs can undoubtedly slow down tumor growth. Most often, the survivors of the chemical attack come alive and reignite the cancer's fire. Cytotoxic drugs also damage normal cells due to the lack of selectivity of chemical agents (as opposed, for example, to monoclonal antibodies). The fast-growing cells in the mouth and digestive tract are particularly susceptible to damage, resulting in common side effects such as ulcers in the mouth (and throat) and gastrointestinal disturbances accompanied by extreme nausea. The hematopoietic stem cells that give rise to the cells of the blood and immune system are also highly susceptible and, when damaged, anemia and increased susceptibility to infection can result. The damage can be so severe that it spawns a new cancer.

Add to the list of common chemotherapeutic side effects neuropathies (nerve damage), heart and lung abnormalities, kidney disease, and damage to the reproductive organs, and it becomes clear why these innovative approaches to cancer treatment offer the hope of a better patient experience. We know that the more we enable alternatives to treatment beyond toxic chemical agents and replace these chemical sledgehammers with precision molecular tools, the better the outcomes can be.

This perspective on the future of cancer treatment assumes that better control can be achieved over the major side effects induced by CAR-T and other immunotherapeutic agents (e.g., checkpoint inhibitors, oncolytic viruses, and bifunctional antibodies). Along with their potential to achieve spectacular results, these potent modulators of the immune response are also capable of causing potentially lethal side effects. Currently, the anti-interleukin-6 antibody used by Carl June to save Emily Whitehead from CRS in 2012 (Chapter 13) is typically used to dampen the runaway immune response seen in cytokine release syndrome. The problem of vascular leak syndrome is an area of intensive investigation that remains a significant challenge in clinical care. Progress in controlling these potentially deadly side effects of immunotherapies is needed to proceed toward the vision of the widespread use of immunotherapeutics to treat cancer.

Finally, additional antibody therapeutics (including bi- and even tri-functional forms) that disrupt signaling pathways and/or bring different cell types into proximity with each other to slow tumor growth and/or destroy tumor cells will become part of routine oncological practice in the coming years.

If we have such powerful anti-cancer drugs, should we try to cure all cancers? In the 2019 interview referenced above, Dr. June states, "I think we have to cure it," noting, "If we turned cancer into a chronic disease, it would be unaffordable."[357] So how can we best move forward to improve the accessibility of these and other expensive medicines?

Implementation of these therapies would benefit from a higher success rate. It would not be sustainable to administer CAR-T therapies, for example, on large patient populations if the treatment renders a medically meaningful response for only a small or modest percentage of patients. While an oncology drug with a complete remission rate of 25% is acceptable today, the CR rate must increase significantly to enable widespread patient access to these new treatment methods.

The affordability of healthcare, and the tremendous disparities in healthcare delivery, must be addressed as medical treatments become more sophisticated; otherwise, quality healthcare will become increasingly skewed toward the affluent, and in time, the system may collapse in chaos.

It is likely that the healthcare system in the United States cannot withstand the cost of the widespread use of therapies designed to bring long-term remissions of metastatic cancer and/or tumor eradication for all patients who might benefit. Selecting the patients who are most likely to succeed from the treatment using molecular data and other medical information provides the best "bang for the buck."

One approach, already well underway, is to screen the genomic markers and biochemical characteristics of each cancer to select patients who meet the molecular profile known to favor therapeutic success. Pre-screening patients might increase the success rate of therapy into the 45-50% range, thereby reducing the enormous costs of treating many patients who will not derive benefit from these overly expensive, patient-specific treatments.[358]

The power of immune checkpoint inhibitors to unleash the host immune system is being explored further to improve both the efficacy and the safety of these therapeutics. As there are multiple immune checkpoints, and some that have barely been explored beyond CTLA-4 and PD-1, there are many potential combinations of immune checkpoint inhibitors (alone and with other therapeutics) yet to be evaluated in the search for the most impactful way to empower the immune system.

Currently, there is significant research on leveraging the growing knowledge of the critical processes encompassed by the mechanistic pillars of cancer to develop new therapeutics. One idea under investigation is to turn the genetic instability of cancer cells from an evolutionary benefit for rapidly mutating cancer cells into a fatal flaw. Since most tumor cells have damaged DNA repair mechanisms, interfering with the residual DNA repair pathways would cause so much genetic damage that the genetically unstable tumor cells can be pushed "over the genomic edge" to a point where they can no longer survive. This approach has been successfully applied in the development of the PARP inhibitors.

Rapid advancements in our understanding of the biochemical intricacies of cell differentiation and the families of transcription factors that regulate this intricate process are opening new therapeutic approaches for stopping metastatic spread. As we saw in Chapter 9, a mixture of transcription factors can induce differentiated cells to acquire the characteristics of stem cells. In this process, a less differentiated cell is created from a more differentiated one. If this is possible, why wouldn't it be possible to take a less differentiated (more "stem-like") cell and push it toward a fully differentiated state that does not share the plasticity of cancer cells? The expected outcome would be a diminution of invasive and aggressive behavior by the cancer cells, up to and including the prevention of invasion of the surrounding tissues from the primary tumor. Under such conditions, the likelihood of metastasis would be reduced, perhaps significantly.

Interventions targeting the unique metabolic profile of cancer cells can, in principle, be used to interfere with cancer cell proliferation. The oncogenic signaling that drives cancer progression alters the activity of metabolic enzymes in cancer cells. Significant research is ongoing to interfere with the activity of these enzymes. As an example, many cancer cells over-express the enzymes responsible for glycolysis. Inhibition of these enzymes can slow glycolysis and deprive cancer cells of the metabolic intermediates required for rapid cell proliferation.[359] Additional potential targets include the enzymes responsible for the uptake of glutamine, the major nitrogen source for cell growth, as well as those driving fatty acid metabolism.[360]

Since the metabolic changes found in cancer cells are extraordinarily complex, intensive research is required to develop metabolic interventions. Here is a place to apply the power of artificial intelligence-based systems to analyze copious amounts of data and apply this information to the development of a multipronged (aka *combinatorial*) process for the manipulation of cancer cell metabolism.[361] In this approach, multiple inhibitors of metabolic pathways (such as glycolysis, glutamine metabolism, and the Krebs cycle) could be used in combination to optimize the metabolic starvation of cancer cells while sparing normal cells. The goal is to establish "a metabolism-based precision medicine platform that can best predict personalized drugs, which synergize with a patient's metabolic status."[362]

As the emerging immunotherapies undergo improvements to enhance their success rate in achieving durable remissions, technologies will emerge that enhance therapeutic efficacy. One idea is to use gene therapy to insert genes for intact copies of mutated tumor suppressors into tumor cells to compensate for non-functional, mutated tumor suppressors.

It might also be possible to correct mutations in tumor suppressors using gene editing technology and/or to turn off or delete oncogenes that drive tumors. The 2020 Nobel Prize in Chemistry was awarded to Jennifer Doudna and Emmanuelle Charpentier for the development of CRISPR-Cas9 gene editing. The technology leverages an innate bacterial defense against viruses that can remove and replace DNA nucleotides with astounding specificity.

Gene editing technology offers the intriguing possibility of long-term genetic corrections of human DNA. The emergence of this method for editing genes is an astounding achievement, the first time such a capability has been available that can alter the genetic code in living organisms. An ancient biochemical mechanism, operational in microbial systems for eons, now provides a tool that can repair damaged genes and to selectively turn individual genes on and off.

Gene editing is a revolutionary technology that forever alters the paradigm of molecular genetics, supercharging the field of cellular engineering from the level of the gene down to the level of individual nucleotides. Like any significant emerging technology, the prospect of its use carries significant downside risks. Our understanding of the staggeringly complex ballet of gene expression is far from complete.

The prospect of altering the body's germ cells obviously tops the list of potential risks for this technology. Since these cells transfer genetic information to future generations, changing them can permanently alter the human gene pool. To most scientists who understand the hazards of fiddling with the gene pool, editing the genes in the germ cells seems like a bad idea.

Changing the genome of the body's somatic cells carries far lower risk than changing the human gene pool. Nonetheless, the genetic code features layer upon layer of intricate coding. Even a precise change in a single nucleotide to correct a genetic defect in a protein carries uncertainty and risk. This is because the same section of the genome that codes for the protein under repair might be expressed in other contexts, such as microRNAs with coding sequences overlapping those of a protein.

Some microRNAs serve functions critical to homeostatic controls over proliferation, metabolism, angiogenesis, and differentiation. Disruption of these processes can have devastating effects on cells, including the promotion of tumor growth. In addition, the presence of three different reading frames in the code raises the possibility that the altered nucleotide in the edited sequence that corrects a defective protein might also code for a microRNA or another protein in a different reading frame that a gene edit can disrupt.

Returning to the risks of using gene editing technology to alterations of the germline, it should be noted that the last time the scientific world (and the world at large) faced such consequential decisions regarding a biological technology was a half-century ago. The introduction of recombinant DNA technology in the 1970s provided a method for dicing and slicing genes and assembling the pieces into new genetic constructs. When these constructs are placed into living cells, they can express proteins of interest never before available from natural sources in quantities sufficient for detailed study.

Some argued at the time that since these genetic constructs made in the laboratory never existed during our evolution, we might be tinkering with our very essence in unpredictable ways. For them, concerns ranged from the reasonable possibility of releasing an unrelenting and untreatable pathogen upon the masses, an *Andromeda Strain* on steroids, to ridiculous fears of a new and more formidable Frankenstein's monster. Others argued that over evolutionary history, so many combinations of genetic sequences have existed that it is unlikely humanity can create anything more dangerous than those produced in the natural world.

A series of global scientific meetings in the early to mid-1970s resulted in provisions that have thus far prevented any of the catastrophes some feared at the beginning of the recombinant DNA era.[363] Recombinant DNA technology and the biotechnology industry it spawned have vastly enhanced our understanding of human biology and the mechanisms of disease. At the same time, valid concerns remain about the engineering of crops and the potential impact of these activities on the food supply and the environment.

As gene editing technology matures and the work goes on to create highly accurate gene editing methodologies, ethical questions remain. The biological risks have prompted ongoing debate. There is obvious appeal in a technology that can correct genetic errors at the root of human diseases. Given the medical promise of gene editing, the controversial applications—including the long-debated notion of "designer babies"—will require continued discussion and debate.

Currently, the use of gene editing for the treatment of viable embryos is not sanctioned by the global scientific community, with a consensus that this application is exceedingly dangerous. However, experiments in China in 2018 on non-viable embryos also led to a firestorm in scientific and government circles highlighting the risks of tampering with the human genome.[364]

Further refinements of the immunotherapeutic approaches described in the previous chapters are already under development. The goal is to enhance the properties already known to correlate with therapeutic efficacy. CAR-T cell research is focused on improving the persistence of the engineered cells in the body, reducing toxicity, optimizing co-stimulation pathways, and fine-tuning the binding characteristics of the CAR-T construct to provide the greatest efficacy with the fewest side effects.[365] In addition, research is underway aimed at adding the capability to target multiple antigens to avoid tumor escape through antigen loss when targeting a single antigen on the tumor cells.

Additional genes can be added to the CAR-T constructs, including genes for cytokines that enhance the adaptive and innate immune responses. Gene editing technology can also customize T cells as therapeutic agents. For example, TCR-T cells with inactivated HLA genes would be able to bind tumor antigens via the TCR without HLA restriction.

To reduce the toxicity of CAR-T therapy in cases where the engineered

T cells turn into destroyers of normal tissues, constructs for the CAR-T cells can be created that include a "suicide gene" to create a "suicide switch" that can turn off the activity of the engineered T cells. In the event of unacceptable toxicity, the CAR-T cells can be suppressed by treating the patient with a chemical agent that activates the expression of the suicide gene.

The reversal of T cell exhaustion has attracted significant interest in the cancer research community. In addition to the profound dampening of T cell function due to the prolonged antigenic stimulation found in cancer patients, tumor cells often over-express immune modulatory proteins. This includes proteins that block T cell co-stimulation or activate immune checkpoints. For example, many tumor cells over-express PD-L1 to occupy the PD-1 binding site on T cells, thereby maintaining that immune checkpoint. By so doing, those dastardly tumor cells are doing the opposite of what immune checkpoint inhibitors are designed to do. Tumors also over-express other surface proteins that manipulate the immune response in their favor. One example is *CD47*, a protein currently in the crosshairs of cancer researchers that serves as a "don't eat me" signal that blocks the phagocytosis of normal cells.[366]

Ongoing research has already provided evidence that functional, transcriptional, and epigenetic changes occur in CAR-T cells that are continually stimulated by exposure to their target antigen. Drugs that dampen CAR-T signaling for short periods during treatment are already under development with the goal of overcoming the functional exhaustion of CAR-T cells. The same phenomenon also happens with TCR-T cells, and the same approach to overcoming the exhaustion of CAR-T cells also applies to TCR-T cells.

Finally, significant efforts are underway to create off-the-shelf CAR-T cell therapies using donor (instead of patient) T cells. These allogeneic T cells (the Greek word *allos* means "different" or "other") are engineered to avoid immune rejection in the patient as well HLA restriction of T cell antigen recognition. HLA restriction can also (theoretically) be overcome by using gene editing technology to inactivate the HLA genes of TCR-T cells. If HLA restriction can be eliminated, it is possible to create a TCR-based

T cell therapy that can be used in any patient whose tumor contains the antigen target(s) for which the TCR-T cells are designed. This approach would both reduce the time from diagnosis to treatment as well as reduce the costs of this expensive therapy.

Concurrent with the explosive advances in cancer treatment that have characterized this new century, diagnostic technologies are far beyond what was available in the twentieth century. The same advancements in analytical technology that enable increasingly detailed and sophisticated research in cancer biology and the development of astounding new cancer therapeutics are also being applied in the diagnostic arena. This effort is generating extremely sensitive and reliable methods to assess patient samples for the presence of active cancer-related genes.

In the past decade, rapid advancements in the sensitivity and accuracy of DNA sequencing have enabled the detection of circulating fragments of DNA derived from dead or dying tumor cells. These pieces of DNA, called *circulating tumor DNA* (*ctDNA*), can be identified in the blood.[367] The evaluation of tumors based on what is found in the blood is called a *liquid biopsy*. Where feasible, liquid biopsies avoid surgical excision of a tumor sample compared to a traditional biopsy), which is an invasive procedure carrying inherent risks. The earlier these DNA fragments are detected, the better, and in many cases, even early cancers that are not yet evident using other diagnostic methods can be detected using this approach.[368]

The ability to detect circulating tumor DNA is a true game changer in cancer medicine. This breakthrough provides a molecular signature of cancer across its life cycle, from the time of the emerging *neoplasm* (a fancy word for cancer that literally means "new growth") all the way through the progression to metastatic disease. While early cancers are being detected using this technique today, ongoing developments will improve the accuracy and sensitivity of this diagnostic method, thereby providing a readily accessible means of monitoring tumor status that has never been available to medical science.[369]

While successful excisions of early tumors that have yet to spread their deadly tendrils throughout the body can lead to tumor eradication, achieving such a result for cancer patients who have entered the metastatic stage is not feasible. Historically, the goal of therapy for advanced cancer has been to contain tumor growth and, hopefully, to slow its progress. Regardless of the mode of treatment, these tumors usually recur over time and are usually related to the cause of patient death.

The cancer therapy of the future (by the middle of this century, or thereabouts) will differ in significant ways from the historical and current treatment regimens. Cancer diagnosis and treatment will be based on the molecular characteristics of the cancer itself rather than on its anatomical location in the body and visual appearance in the microscope. No longer will we merely use the anatomical location ("lung cancer," "colon cancer," etc.) to identify a tumor. Rather, each individual cancer will be defined by its driver mutations and other molecular characteristics. This approach will facilitate a therapeutic response fine-tuned to manage the biochemical changes induced by the cancer.[370]

Typically, treating metastatic cancer requires administering one or more therapeutic agents followed by an observation period and subsequent diagnostic scans (e.g., MRI and CT scans) to determine the tumor size and appearance before initiating the next therapeutic regimen. The capability to continuously monitor the genetic and biochemical fingerprint of the tumor population using ctDNA, mass spectrometry, and other sophisticated analytical techniques offers an objective means to understand the effectiveness, or lack thereof, of each round of treatment.[371] The information on the molecular characteristics of the tumor will provide a tremendous level of flexibility, including responses in real-time to ineffective or deleterious treatments.

Although we are only in the early days of the immunotherapy revolution, dozens of clinical trials are already looking for ways to combine immunotherapeutic agents with other immunotherapies and/or targeted drugs. One such approach under evaluation is using oncolytic viruses to release tumor antigens into the circulation along with checkpoint inhibitors that remove the brakes from immune cells. Similarly, combining cancer vaccines with oncolytic viruses can enhance the vaccine's effects by releasing tumor antigens from tumor cells. In addition, checkpoint inhibitors are under evaluation in combination with CAR-T therapies and ADCs to amplify the immune response relative to that obtained with a single immunotherapeutic agent.

Many obstacles remain to accomplish the dream of personalized medicine. We should be under no illusion that the path will be easy or free of setbacks. There are still significant barriers to success in the effort to extend the gains made in liquid cancers to the treatment of solid tumors, which comprise about three-quarters of all human cancers.[372] Early attempts at CAR-T therapy for the treatment of solid tumors have demonstrated poor efficacy with significant toxic effects. Recent data showed (short-lived) efficacy of the engineered T cells with acceptable toxicity in several cancers, including difficult-to-treat cancers such as pancreatic cancer.[373]

There is breaking news on the CAR-T front that demonstrates the promise of CAR-T cell therapy. In an early clinical trial at Mass General Cancer Center, CAR-T cells combined with a bifunctional antibody demonstrated astonishing efficacy in glioma patients. Glioma is the most abundant brain cancer, with a dismal prognosis and few therapeutic options. Amazingly, the three patients in the trial all showed significant tumor regression, with one patient, a 57-year-old female, experiencing what was termed "a near-complete tumor regression."[374] Now get this: these results were obtained within a week of a single treatment. If such results are possible in such a deadly and difficult-to-treat cancer, the future for combination immunotherapy is bright indeed.

The treatment of melanoma with TILs demonstrates that solid tumors can be successfully treated with T cell-based therapies under the appropriate circumstances. While data from clinical trials have shown that engineered T cells, once present in the bloodstream, can reach the tumor(s) "but fail to expand, persist, and mediate objective responses."[375] The key will be to transfer the astounding successes in the treatment of liquid cancers—including instances where the CAR-T cells persisted in patients for periods as long as a decade—to solid tumors.

There is also breaking news on this front: on February 16, 2024, the FDA approved the first cell therapy treatment for solid tumors. Amtagvi®, which uses the tumor-infiltrating lymphocyte technology pioneered by Steven Rosenberg, is now available for the treatment of melanoma patients.

Additional technological advancement is needed in multiple areas to extend effective cell-based therapies to solid tumors. These developments will leverage our increasing understanding of the TME as a highly dynamic environment that directly impacts the course of tumor progression, evolving along with the tumor. Like the tumor cells themselves, the cells in the TME evolve in response to changing conditions, including those introduced by therapeutic intervention. By evolving along with the successful cancer clones, the TME supports therapeutic resistance following treatment.

Modulation of the immunosuppressive TME is a top priority of the current efforts to expand immunotherapies beyond liquid cancers. The goal is to alter the characteristics of the TME to create an environment favorable to the activation and persistence of cytotoxic $CD8^+$ and helper $CD4^+$ T cells, thereby enhancing antigen presentation and promoting effective collaboration between the immune system's adaptive and innate arms.

Given the accelerating pace of discovery in cancer biology and tumor immunity, it is now feasible to use monoclonal antibodies, engineered cells, and oncolytic viruses to modulate the TME to favor tumor destruction. In addition, a molecular understanding of the differences in the TME between responders and non-responders to cancer therapeutics (including immune checkpoint inhibitors) would provide valuable information on the alterations of the TME that might induce therapeutic sensitivity in the tumor cells of non-responders.

Based on our current understanding of the dynamics of the TME, there is considerable interest and research activity on several other fronts, including efforts to reprogram the TME by reducing the activity of immunosuppressive cells, awakening exhausted T cells, increasing the cytotoxicity of $CD8^+$ cells, enhancing the functionality of $CD4^+$ helper cells, and stimulating innate immunity to render a potent anti-tumor response.[376]

Increased molecular understanding of the dynamics of the TME and the factors that impact treatment outcomes (including drug resistance) could potentially enable treatment regimens that stay a step ahead of resistance and thwart the tumor's attempt to adapt. In addition, innovative technologies in drug delivery, some already under development, will provide new ways to concentrate the effects of therapeutic agents where they are most needed.[377]

In addition to the 30 trillion (or so) cells present in our bodies, our organs are both covered with, and occupied by, trillions of microorganisms.[378] These organisms cover a vast range of microbial life, including bacteria, viruses, fungi (e.g., yeast), protozoans (e.g., amoeba), bacteriophages (viruses that infect bacteria), and *archaea*, organisms that were formerly classified as bacteria. The collection of organisms found throughout our bodies is called the *microbiome*.

When we look at the number and variety of microorganisms that accompany us on our journey through life, we recognize there are at least as many of them as there are of us. Perhaps more. Estimates of the number of microorganisms in our microbiomes range from about the same number of microbes as human cells (about 30 trillion) to as much as one (and perhaps two) orders of magnitude more microbes than human cells (an astounding 300 trillion to 3 quadrillion organisms).[379]

Approximately four decades ago, the archaea were split out into a separate category of life based on their unique biological properties. Archaea bear some similarities with prokaryotes, single-celled organisms such as bacteria that lack a nuclear membrane and cytoplasmic organelles. In other respects, archaea are more like eukaryotes, organisms comprised of cells having a nucleus and well-defined organelles. Finally, the archaea have some unique properties, including the compositions of their cell wall and cell membrane, which differ biochemically from those found in the prokaryotes and eukaryotes.

The archaea thus represent a third domain (type) of living organism that is not a prokaryote or a eukaryote but something altogether different. First discovered in extreme environments, such as at high and low temperatures, under extreme pressures, and in highly acidic conditions (such organisms are called *extremophiles*), we have only recently come to understand that archaea are not only found in nature: they are also inside of us. Archae comprise a tiny fraction of the microbiome, about 1.2%, according to one source,[380] or as low as 0.1%, according to another.[381] In relative terms, we know little about these fascinating microorganisms.

The microbiome is a highly dynamic system in constant communication with the cells of host immunity. Over the lifespan, our immune systems learn to discriminate the harmless and beneficial (collectively known as *commensal*) organisms from those posing a threat (*pathogenic organisms*), thereby inducing tolerance to the former while mounting a vigorous response to the latter. As a result, the microbiome impacts health and disease in profound ways, with effects on the regulation of inflammation, metabolism, and host immunity.[382]

Maintaining homeostasis in the microbiome depends on the outcome of the competition for resources between the diverse types of microbial organisms resident in the body that provides a balanced environment where the microbiome contains the appropriate variety and amounts of organisms in the absence of significant numbers of pathological strains that can compromise homeostasis. In this battle within us, some microbes create toxic and damaging molecules with the potential to impact competing species. Such molecules can, for example, damage DNA, impact metabolic balance, and interfere with cellular signaling in other microbial species.

Under homeostatic conditions, these antimicrobials in the microbiome have minimal impact on at least one species: the human host. Under non-homeostatic conditions, where the microbiome's components are no longer in balance (a state called *dysbiosis*), the human host can be severely compromised if organisms capable of interfering with our biochemical pathways and disrupting immune regulation come to dominate the microbiome. Such organisms are not necessarily pathogenic under homeostatic conditions; the overgrowth of an ordinarily commensal organism can contribute to dysbiosis. Dysbiosis can cause significant (and perhaps even grievous) harm, including impacts on cell proliferation and differentiation (stemness) that are linked to the progression of cancer.[383]

While only a handful of bacteria are recognized as human carcinogens (these are called *oncomicrobes*),[384] other microbes (called *complicit microbes*) that do not directly cause cancer participate in cancer promotion by modulating immune function (for example, by secreting molecules that enhance the biosynthesis of pro-inflammatory cytokines by host immune cells) and generating, as noted above, pro-tumorigenic metabolites (products of microbial metabolism) that impact homeostatic functions.[385] Strangely, some human tumors depend on the presence of microbial metabolites. As an example, the carcinogenicity of many p53 mutations is dependent on the presence of a microbially-produced metabolite called gallic acid.[386]

While most of the microbes that comprise the microbiome are found in the intestines, bacteria are also found within tissues, where their metabolic byproducts interact with host biochemistry and host immunity.[387] For reasons that are not yet clear, some bacteria have a natural affinity for cancer cells. This property is being exploited by research aimed at creating engineered bacteria that can serve as potent therapeutic agents that accumulate in the tumor bed and release cytotoxic and/or immunomodulatory payloads, serving as "mini bioreactors" that continuously generate molecules with anti-tumor properties.

While the role of the microbiome in tumor progression and treatment response is complex, recent data demonstrate that specific microbes are favorable[388] (and others are unfavorable)[389] in the context of cancer treatment. The presence of specific commensals has been associated with anti-tumor T cell responses. In addition, the presence in the microbiome of certain species of the bacterial genus *Bifidobacteria* correlates with enhanced responses to anti-PD-1 immune checkpoint inhibitor antibodies. Finally, the presence of several specific commensals in the gut microbiome enhances the activity of antigen-presenting dendritic cells that can stimulate anti-tumor T cell responses.[390]

The composition of the microbiome directly impacts the tumor microenvironment by influencing the infiltration of immune cells into the tumor bed.[391] For these reasons, active research is ongoing on the use of prebiotics and probiotics.[392] In addition, there is growing interest in *fecal microbiota transplantation* (FMT), in which fecal matter from responders to therapy is transplanted into the intestines of cancer patients.

Unlike for liquid tumors, delivery of therapeutic agents to solid tumors is hampered by the unfavorable conditions of the chaotic, pro-inflammatory TME. In addition, effective therapeutics need to breach the physical barrier to drug penetration into the tumor presented by the ECM and the stromal cells. This might be achieved by the targeted delivery of the appropriate matrix metalloproteinases to break down the ECM along with highly targeted radiation therapy, which effectively reduces the stromal cell population supporting tumor growth. Finally, monoclonal antibodies can be used to target tumors by incorporating them in exosomes to deliver payloads that modulate the TME and directly attack cancer cells from within.

There is no question that lifestyle choices are critically important when it comes to tumor formation and progression. We know from experience (and common sense) that smoking, the excessive consumption of processed sugar, alcohol, and red meat, as well as obesity, are detrimental to our health. These factors contribute to the creation of inflammatory environments favorable for tumor growth (as well as type 2 diabetes, heart disease, and other inflammation-driven diseases). We know that we tip the odds in our favor by doing what we can to support a healthy immune system to protect us from both infection and cellular degradation. We also know that environmental pollution, including airborne particulates, microplastics, and toxic industrial by-products, are now a planet-wide concern due to their detrimental effects on our health as well as the health of our planet.

Nutrition is paramount. So is daily exercise. Easier to say, but much harder to do, in face of the challenges of modern life. We live today immersed in the inordinate stresses of twenty-first century living, amidst vast economic inequality and uncertainty. In an ideal world, we would all be able to keep our bodies supplied with a proper balance of nutrients. The reality is that healthy calories from fresh, high-quality foods are far more expensive than the calories provided by the vast array of highly processed foodstuffs that are nutrient-deficient and calorie-dense. There must be a better way.

There are some simple things we can do. As we are mostly water, it is critical that we stay hydrated. We should keep moving our bodies as best we can, protect our skin from the sun and other forms of ionizing radiation, and get adequate rest. As noted above, we all know about the importance of proper nutrition, that we "are what we eat." This can present some challenges: as a newcomer to Virginia, I admit it is exceedingly difficult to resist the aroma wafting out of the nearby fried chicken shack. We have indulged from time to time, and it is admittedly delicious. There is wisdom in moderation here, as in all things, including vitamins and other supplements. There is such a thing as too much, even when something is good for us in smaller amounts.

Cancer screening is a critical tool that we ignore at our own peril. A knowledge of family history goes a long way here. While most cancers are not hereditary, similarities in biology between us and our relatives might be informative, if not predictive. We can also benefit from staying in touch with our bodies (yoga and meditation excel here), so that we recognize when something seems "off." If so, it is best not to ignore it. Usually there will be a mundane explanation; sometimes, unfortunately, not so much.[393]

Self-advocacy may be required. Seek information from true expert sources. While there are many opinions out there, some are more useful and informed than others. Work with your most trusted healthcare provider(s) to help you find answers. If you are told that the odds are low that you have cancer, and you are unconvinced, remember that statistics apply to populations and not individuals. There are always outliers.

Also important, but more challenging, is to do what we can to reduce our stress and anxiety. It is no coincidence that we are more vulnerable to illness when we are overwhelmed. It saps our strength, disturbs our sleep, and impedes immune function. This is when we come down with colds or succumb to the return of symptoms of the chronic ailments that plague us. Untreated anxiety also degrades our ability to resist disease. Our immune cells, which have receptors for neurotransmitters, are in direct communication with the nervous system. They feel our pain.

We now stand at the precipice of a new age in cancer treatment, with our eyes on a future where a diagnosis of metastatic cancer need not be a harbinger of intense suffering and inevitable death. Seven decades of a revolution in biology wrought by the discovery of the structure of DNA have taken us from a rudimentary understanding of biochemistry, immunology, and heredity to a new paradigm in the treatment of human ailments, including cancer, informed by molecular information that can be leveraged to attack the root causes of human diseases.

The new and evolving methods of cancer treatment described in the previous chapters offer a message of hope to current and future cancer sufferers. Even today, these astounding, novel approaches are eradicating cancers in patients with advanced leukemias and lymphomas. Tomorrow, these advances will be expanded to cover more patients with not just liquid cancers but also with solid tumors.

For other patients who are not eligible for treatments with the potential for tumor eradication, our increased understanding of the biochemical mechanisms at the heart of cancer progression offers the possibility their cancers can be managed for prolonged periods, perhaps even decades, with a far higher quality of life than that experienced today by patients with metastatic cancer. For cancer researchers, physicians, and, most importantly, patients and their families, that would indeed be sweet.

Notes

References for journal articles can be found online at the National Library of Medicine's website pubmed.gov. To find an article, you can type the name into the search bar or list one or more authors, last name followed by initials. There will minimally be an abstract available for each reference, and in many cases, the entire manuscript is available if you see the link "Free PMC Article."

Chapter 1
[1] D Masopust, V Vezys, EJ Wherry, and R Ahmed, A brief history of T cells. *Eur J Immunol* 37: S103–110 (2007).
[2] The final requirement is to isolate the organism from the newly infected individual and successfully grow it in pure culture that is unadulterated by other infectious agents.
[3] S Reidel, Edward Jenner and the history of smallpox and vaccination. *BUMC Proceedings* 18:21–25 (2005).
[4] ibid.
[5] P Ehrlich, Partial Cell Functions. *Nobel Prize Lecture*, December 11, 1908.
[6] ibid.
[7] ibid.
[8] K Strebhardt and A Ullrich, Paul Ehrlich's magic bullet concept: 100 years of progress. *Nature Rev Cancer* 8: 473-80 (2004).
[9] Racine, Valerie, Ilya Ilyich Mechnikov (Elie Mechnikov) (1845-1916). *Embryo Project Encyclopedia* (2014-07-05). ISSN: 1940-5030 http://embryo.asu.edu/handle/10776/8018.
[10] ibid.
[11] SY Tan and MK Dee, Elie Mechnikov (1845-1916): discoverer of phagocytosis. *Singapore Med J* 50(5): 456-7 (2009).

[12] ibid.
[13] ibid.
[14] ibid.
[15] II Mechnikov, On the present state of the question of immunity in infectious diseases. *Nobel Prize Lecture*, December 11, 1908.
[16] Upon removal of all the cells from the blood, the remaining liquid, called plasma, contains proteins that participate in the clotting of blood. In this protein mixture we find the critical clotting protein Factor VIII (factor 8), a deficiency of which causes the most abundant form of hemophilia (Hemophilia A). The clotting proteins are removed from the plasma to create serum.
[17] The other humors were black bile, yellow bile, and phlegm.
[18] The requirement for training of the adaptive response explains why newborns have a limited capability to respond to antigens, and why breastfeeding infants is an important means for instilling immunity via maternal antibodies that are passed from the milk to the infant.
[19] Herein we find one of the issues with using antibody levels in assessing the duration of the immune response after vaccination. The more time that passes after the first antigen exposure (priming), the fewer antigen-specific antibodies we find in the blood due to decay of the activated B plasma cells. This measurement tells us nothing about the status of the B memory cells, which are found in the lymph nodes after antigen priming in both B and T cell lineages.
[20] Figure 2 depicts human IgG1, which has two disulfide bridges in the hinge. IgG2 has a more complex disulfide bond structure, with four disulfide bridges. Most human antibody therapeutics are members of either the IgG1 or IgG2 sub-class (IgG3 has 11 hinge disulfides, and IgG4 has 2).
[21] Amino acid sequences in the C_H2 domains of IgGs contain binding sites that mediate the antibody "half-life"—a measure of how long the antibody remains in the circulation—as well as sequences that interact with protein receptors on T cells, macrophages, and other immune cells that participate

in the immune response. The effector functions are elicited by interactions between antibody-antigen complexes and receptors on immune cells. Activation of these effector functions, called antibody-dependent cellular cytotoxicity (ADCC) and antibody-dependent cellular phagocytosis (ADCP), results in the death of targeted cells by cytotoxic (killer) T cells (in ADCC) and activated phagocytic cells, such as macrophages (in ADCP). The third type of effector function, called *complement activation*, involves the interactions of complement proteins in the blood with the Fc domains of antibody-antigen complexes at cell surfaces. These interactions result in the creation of a *membrane attack complex* that punctures the membrane and destroys the cell. The complement system is a critical component of innate immunity. As noted in the text, the FcRn receptor is a "salvage" receptor that prevents rapid degradation of antibodies and thereby prolongs their residence time in the body. The lengthy residence time of antibodies explains (at least in part) why they have been so successful as therapeutic agents.

[22] C Milstein, From the Structure of Antibodies to the Diversification of the Immune Response. *Nobel Lecture*, 8 December 1984.

[23] The degree was Milstein's second biochemistry Ph.D. His first Ph.D. in biochemistry was obtained in 1957 at the University of Buenos Aires. (I, on the other hand, found that one Ph.D. was quite sufficient.)

[24] The analytical method used to detect antibodies that bind to the target protein is called an Enzyme-Linked Immunosorbant Assay (ELISA)—or, more simply, an immunoassay—a method that has been a workhorse in biochemistry and immunology laboratories for over half a century.

[25] The experimental details can be found in G Köhler and C Milstein, Continuous cultures of fused cells secreting antibody of predefined specificity. *Nature* 256, 495-497 (1975).

[26] Milstein's Nobel Lecture was entitled "From the structure of antibodies to the diversification of the immune response." The 1984 Nobel Prize in Physiology or Medicine was awarded in recognition of both the

monoclonal antibody technology invented by Kohler and Milstein, as well as the important discoveries about immune diversity and specificity by Niels Jerne.

[27] A Vassilev and ML DePamphilis, Links between DNA Replication, Stem Cells and Cancer. *Genes* 8 (2): 45-78 (2017).

[28] ibid.

[29] The antigen size can vary. If we assume an average of seven amino acids as the size of the typical antigen, there are 20^7 sequence possibilities, or 1.28 billion possible combinations.

Chapter 2

[30] C Raup, Teratomas: The Embryo Project Encyclopedia (2010), *embryo.asu.edu/pages/teratomas*

[31] LC Stevens, Jr. and CC Little, Spontaneous Testicular Teratomas in an Inbred Strain of Mice. *Proceedings of the National Academy of Sciences of the USA* 40: 1080-7 (1954).

[32] ibid.

[33] ibid.

[34] R Lewis, A Stem Cell Legacy: Leroy Stevens. *The Scientist* March 6, 2000.

[35] ML Suva, N Riggi and BE Bernstein, Epigenetic Reprogramming in Cancer. *Science* 339 (29 March 2013), 1567-70.

[36] *www.merriam-webster.com/dictionary/plasticity*

[37] JD Watson and FHC Crick, A structure for deoxyribose nucleic acid. *Nature* 171, 737–738 (1953).

Chapter 3

[38] The six atoms of macromolecular life, known by the acronym "CHNOPS," are carbon, hydrogen, nitrogen, oxygen, phosphorus, and sulfur. These atoms, along with trace elements such as metal ions like magnesium, manganese, iron, and others, provide the ingredients for life on Earth.

[39] Insulin, a small protein, has a molecular mass of 5808 Daltons. Human serum albumin, the most common protein in our blood, has a molecular mass of about 66,500 Daltons, and our antibodies, which are comprised of 18,000 - 20,000 atoms, have a molecular mass of about 145,000 Daltons.

[40] Electrons are shared in pairs in covalent bonds. The sharing of a single pair of electrons forms a single bond; the sharing of two pairs results in a double bond, and, you guessed it, the sharing of three electron pairs results in a triple bond, the strongest type of chemical bond possible.

[41] The hydrogen atom involved in hydrogen bonding is itself covalently bound to an electronegative atom, that is, an atom with a strong attraction to electrons. A hydrogen atom that is bonded to an electronegative atom has a partial positive charge because the electrons in the electromagnetic atom tend to compress toward the nucleus. Such atoms create what is called a "dipole" in which the charge is unequally distributed in space. The partially charged hydrogen atom interacts via charge attraction with the negatively charged region of another dipole (another electronegative atom). Such interactions, which are abundant in proteins and DNA, are found within a molecule as well as between molecules (e.g., the hydrogen bonds between nucleotides on the opposing strands of the DNA double helix).

[42] Typical conditions might be 6 molar hydrochloric acid (about 50% HCl) in a sealed glass ampoule held at 115°C (239°F) for 18 hours. At least that was how we did it in 1980. Modern amino acid analysis methods leverage microwave radiation for a faster and more reproducible digest.

[43] The order of the amino acids in the chain is commonly called the amino acid sequence; it is also known to biochemists more formally as the primary sequence. Biochemists call the individual amino acids "residues." As an example, if the fifth amino acid in the chain is a glycine, we say there is a glycine residue at position 5.

[44] Protein function does not work in an on/off manner like an electronic circuit. It is more like a rheostat, with many possible gradations of functionality, depending on the type and extent of damage to the protein

structure and the surrounding conditions. Small perturbations in structure might (or might not) reduce the functionality of the protein, whereas larger structural changes might significantly reduce or even eliminate the biological activity of the protein.

[45] Recent work using artificial intelligence is closing the gap in our knowledge of protein folding using powerful predictive algorithms. We stand on the threshold of the long-sought capability to accurately predict the folding of proteins.

[46] There is no need to go out and load up on antioxidants. In fact, it can be harmful. In excess, antioxidants can cause oxidation (they become pro-oxidants). In general, the rules of moderation apply in dietary supplementation as they do in what we consume at meals. There are special circumstances (vitamin D deficiency is an example) where high doses might be needed to bring a patient back into normal range.

[47] The hydrogen peroxide formed by SOD is converted to water by the enzyme catalase. Superoxide dismutase was discovered by biochemist Irwin Fridovich and his graduate student Joe M. McCord in a laboratory down the hall from my research group at Duke University nine years before my matriculation in the Biochemistry graduate program.

[48] The most oxidized amino acid in proteins is methionine, followed by cysteine and tryptophan.

[49] NJ Curtin, DNA repair dysregulation from cancer driver to therapeutic target. *Nature Rev Cancer* 12: 801-17, 2012.

[50] A damaged protein can be tagged biochemically by enzymes that add a molecular "flag" that marks the defective protein for destruction, leading it to a location in the cell where it is chewed up into amino acids by an enzyme called a protease, followed by recycling of the undamaged amino acids into the synthesis of new proteins.

[51] The word apoptosis is derived from the Greek: *apo*, meaning "from," and *ptosis*, meaning "falling," combined to provide a word meaning "falling off."

[52] S Mukherjee, *The Emperor of All Maladies: A Biography of Cancer*. (New York: Scribner, 2010), p 367. Dr. Mukherjee's wonderful book lucidly presents the fascinating history of cancer and cancer treatment from antiquity through the early years of the 21st century. The major themes for what became *Taming Cancer* began their incubation shortly after I read *Emperor*.

[53] Francis Crick was a trained physicist who helped develop ingenious mines and mine-detection devices for the Royal Navy during World War II. After the war he turned his attention to the emerging field of structural biology and began a path to an appointment with destiny at the Cavendish Laboratory at Cambridge University. There, Crick and the young American biochemist James Watson worked out the DNA structure. Other physicists at the time also began to apply their scientific skills to biological problems, including Ernst Schrödinger and Oxford University biophysicist Maurice Wilkens. Wilken shared the 1962 Nobel Prize in Physiology or Medicine with Watson and Crick for the discovery of the DNA structure. It was in Wilkens' lab where Rosalind Franklin analyzed crystalline DNA by bombarding it with X-rays using a technique called X-ray crystallography, which was also used, most notably by Linus Pauling at Cal Tech, to work out the structural features of proteins. There has long been a controversy about why Rosalind Franklin was not also awarded the Nobel Prize for the DNA discovery. The answer is found in Nobel protocol, as all recipients must be alive at the time of the award, and Dr. Franklin died tragically at age 37 in 1958. While some contend that, as a woman, she would not have been included in the award had she lived, as an eternal optimist I like to think that she would have been a prize recipient for her work in obtaining her famous "photo 51" that provided the evidence for the double helical structure of DNA. As another manner of protocol, the prizes are limited to three recipients. I still choose to believe that she would have received the honor instead of her laboratory head, Dr. Wilkens, for obtaining data that forever altered the study of biology. We'll never know.

[54] A Knudson, Mutation and cancer: statistical study of retinoblastoma.

Proceedings of the National Academy of Sciences of the United States of America 68 (4): 820–3, 1971.

[55] ibid.

[56] ibid.

[57] JJ Yunis and N Ramsey, Retinoblastoma and subband deletion of chromosome 1d3. *Am J Disab Child* 132:161-3, 1979.

[58] Examples of agents that can be used to induce cell hybridization include the addition of calcium chloride to the media, as well as a technique called electroporation, in which an electric current is used to produce tiny pores in the cell membrane.

[59] The loss of the intact copy of the gene (a single gene copy is called an allele) is called a loss of heterozygosity (LOH, in the parlance of genetics).

[60] RA Weinberg, Oncogenes, Antioncogenes and the Molecular Basis of Multistep Carcinogenesis. Cancer Res 49: 3713-21 (1989).

[61] Our culture is laden with militaristic expressions. During my career, major manufacturing and product development issues were countered by meetings in the "War Room," an ironic name for a healthcare company intent on alleviating human suffering.

[62] Milestone (1971): President Nixon declares war on cancer, *dtp.cancer.gov/timeline/noflash/milestones/M4_Nixon.htm*

[63] In time, cancer researchers would realize that only a small percentage of human cancers are caused by viruses, perhaps about 15%. Nonetheless, the focus on cancer-inducing viruses led to the discovery of oncogenes—genes that, when activated by mutation, can lead to the formation and growth of a tumor.

[64] As I was 16 at the time, "helplessly hoping" to play the guitar like Crosby, Stills, and Nash after seeing the "Woodstock" movie, I don't have personal recollections on the matter.

[65] While infection with Hepatitis C can also induce liver cancer, there is no vaccine available for Hep C. However, an anti-viral cocktail developed by Gilead Sciences (Harvoni)® can eradicate the Hep C virus from the

body, the first and only anti-viral that has ever achieved this remarkable result.

[66] JC Rotondo et al., Association Between Simian Virus 40 and Human Tumors. *Front Oncol* 2019; 9: 670. Published online 2019 Jul 25. doi: 10.3389/fonc.2019.00670

[67] ibid.

[68] M Fischer, Census and evaluation of p53 target genes. *Oncogene* (2017) 36, 3943–3956; doi:10.1038/onc.2016.502; published online 13 March 2017.

[69] AM Goh, CR Coffil and DP Lane. The role of mutant p53 in human cancer. *J Pathol* 223:116-126 (2011).

Chapter 4

[70] Length of a Human DNA Molecule, *The Physics Factbook* (Glenn Elert, Ed.), hypertextbook.com/facts/1998/StevenChen.shtml. The result was re-calculated using the updated estimate of 30 trillion human cells instead of the 10 trillion that was used for the 1998 estimate.

[71] ibid.

[72] The math goes as follows: There are about 2 meters of DNA per cell, so 30 trillion cells have about 60 trillion meters, or 60 billion km, of DNA. The distance from Earth to the sun and back is about 300 million km, so the DNA in the body would be long enough to make the round trip about 200 times. Pluto is about 4 billion miles, or 6.5 billion km, from Earth, so a round trip would be about 13 billion km. The DNA in the body would be long enough to make the round trip 4 times and return to Pluto once, with about 1.5 billion km to spare, enough to do the Earth to sun round trip five more times. If that isn't mind-blowing, I don't know what is.

[73] The acetylation reaction neutralizes the positive charge on the nitrogen atom in a lysine, thereby removing the amino acid's positive charge. This change in charge disrupts the electrostatic interactions between the lysine and a negatively charged phosphate on the DNA.

[74] Lecture given on 19 September 1957 by Francis Crick as part of a Society for Experimental Biology symposium on the *Biological Replication of Macromolecules*, held at University College London. hjournals.plos.org/plosbiology/article?id=10.1371/journal.pbio.2003243

[75] FHC Crick, On protein synthesis. *Symp Soc Exp Biol* 1958; 12:138–163. pmid:13580867

[76] A gene is defined here as a nucleic acid sequence that codes for a specific protein, and this is generally how the term "gene" has been widely used. Later in the narrative, we shall see how this definition is limited and subject to revision considering recent findings that have emerged from the human genome and associated projects, which have unraveled the mechanisms responsible for the organization of the genome and control of the expression of the genetic program.

[77] In X-ray crystallography, a molecule is crystallized form solution to form an organized lattice, and an X-ray beam is shot at the target to produce patterns on film that can be used to deduce the structure using a complex mathematical algorithm called a Fourier transformation.

[78] The multiple codons for each amino acid have the same nucleotide at the first two positions with variation(s) at the third position. This allows for the placement of the correct amino acid at that position in the coded protein sequence even if there is a change (e.g., by mutation) at the third position.

[79] Some have argued that the discovery of retroviruses (like HIV) that make a DNA copy of the RNA genome (using an enzyme called reverse transcriptase) reveals a flaw in the central dogma. However, Crick's model did not specify that the transfer of genetic information was in all cases unidirectional.

[80] The paper was published in an edition of the Brookhaven Symposium on Biology dedicated to the evolution of genetic systems. *Brookhaven Symposium in Biology*, Vol. 23, pp. 366-370, 1972.

[81] How much DNA is in a human being?

askanacademic.com/science/how-much-DNA-is-in-a-human-being-871/

[82] S Ohno, Evolution of Genetic Systems. *Brookhaven Symposium in Biology* 23:366-70 (1972).

[83] ibid.

[84] ibid.

[85] ibid.

[86] I was on the side of the skeptics on the Junk Hypothesis, for the reasons stated in the text. Biological systems are efficient relative to what humans can create.

[87] Watson credited Robert Sinsheimer as the first scientist to raise a serious proposal on sequencing the genome at a small meeting in 1985 at the University of California at Santa Cruz, where Sinsheimer was chancellor.

[88] JD Watson, The Human Genome Project: Past, Present and Future. *Science* 248: 44-48 (1990).

[89] ibid.

[90] ibid.

[91] Gilbert and Sanger shared half of the Nobel Prize in Chemistry in 1980 for their work on DNA sequencing. The remaining half of the prize that year went to Paul Berg of Stanford University for his pioneering work on recombinant DNA.

[92] C Cantor, Orchestrating the Human Genome Project. *Science* 248: 49-51 (1990).

[93] NIH Director James Wyngaarden had asked for $50 million; even visits to the House and Senate Appropriations committees by James Watson and David Baltimore asking for $30 million to begin the massive program did not result in a higher allocation. Taken in the context of the 1961 NASA budget of about $550 million (equivalent to about $4.8 billion today), the initial federal allocation for the genome project might be considered as rather paltry. Perhaps this comparison speaks to the fact that the progress of a rocket program is visible to the public, whereas the results of biological research only become evident to taxpayers a long way down the road. This

is not intended as a knock on the space program, which filled my childhood with endless wonder.

[94] FS Collins et al., New Goals for the U.S. Human Genome Project: 1998-2003. *Science* 282:682-9, 1998.

[95] The most prominent industrial participant was J. Craig Venter's Celera Genomics. For Dr. Venter's perspective on the project, see J. Craig Venter, *A Life Decoded: My Genome - My Life*. New York: Penguin Group (Viking), 2007.

[96] For an excellent narrative description of the genome project, see S Mukherjee, *The Gene: An Intimate History*. New York: Scribner, 2016, pp 306-21.

[97] K Mullis, The Polymerase Chain Reaction. *Nobel Prize in Chemistry Acceptance Speech*, December 8, 1993.

[98] Adding heat increases the kinetic energy of the water molecules, which disrupts the hydrogen bonds between the base pairs.

[99] An exception can be found in microbes called thermophiles, which can live at elevated temperatures.

[100] The vast diversity of evolutionary paths available to species on our planet is well exemplified by the extremophiles. In addition to heat-resistant organisms, there are radiation-resistant microbes called radiophiles (such as those found at the Chernobyl nuclear facility), and organisms that live at extraordinarily low temperatures (cryophiles) or in highly acidic environments (acidophiles). The presence of these extreme forms of life in highly hostile environments offers hope to the exobiology community that alien life forms might exist under more extreme conditions than we imagined in earlier times.

[101] Mullis shared the Prize with British biochemist Michael Smith, who was awarded for inventing a technique for creating targeted gene mutations called site-directed mutagenesis.

[102] Without the polymerase chain reaction, the era of DNA fingerprinting, one of the most important forensic tools ever devised, could not have

happened, nor I suspect, would the CSI franchise be enjoying such a long and successful run.

Chapter 5

[103] E Pennisi, And the Gene Number is...? *Science* 288: 1146-7 (2000).

[104] LW Hillier et al., Genomics in C. elegans: so many genes, such a little worm. *Genome Res* 2005 Dec;15(12):1651-60. The genomic data for *C elegans* indicates the presence of 19,735 protein-coding genes. The nematode has a far greater density of genes relative to humans, as its total genome is about 100 million base pairs, approximately 3.3% of the total size of the human genome.

[105] The ENCODE Project Consortium, The ENCODE (ENCyclopedia of DNA Elements) Project. *Science* 306 (5696), 636-40 (2004).

[106] JR Ecker, Genomics: ENCODE explained. *Nature* Vol 489, 52-3 (2012).

[107] Since different genes are active in different cell types, the Encode project looked at 147 types of cells. The data demonstrated that 80.4% of the genome is involved in at least one RNA- and/or chromatin-associated event amongst the cells tested.

[108] S Brenner, F Jacob and M Meselson, An Unstable Intermediate Carrying Information from Genes to Ribosomes for Protein Synthesis. *Nature* 190: 576–581 (1961).

[109] Despite the various mating permutations we might propose for the hermaphrodites, they cannot fertilize each other.

[110] S Brenner, Nature's Gift to Science. *Nobel Lecture*, December 8, 2002.

[111] The mRNA is called the "sense" sequence, since it is read directly from the DNA, and can be used as a template for protein synthesis. The complementary sequence to the mRNA—we can think of it as the "anti-mRNA"— is known as the anti-sense sequence.

[112] 2 October 2006 Nobel Prize Press Release from nobelprize.org announcing that year's Physiology or Medicine recipients for their discovery

of RNA interference.

[113] As a neophyte graduate student, I chose to study the literature on gamma-interferon as the topic of a paper that I would defend orally in front of a faculty committee in the nerve-wracking right-of-passage called the Preliminary Examination (Prelim, for short). Interferon was a fascinating topic for study in 1979, and appeared to have almost magical properties, such as the ability to bind to viral messenger RNAs and inactivate them. This is the underlying basis of the interferon interference phenomenon.

[114] C Mello, Return to the RNAi World: Rethinking Gene Expression and Evolution. *Nobel Prize Lecture*, Dec. 8, 2006.

[115] RNAi appears to be an ancient, and possibly ubiquitous, biochemical mechanism that has been found not only in animals, but also in plants and microbial life.

[116] S Diederichs and DA Haber, Dual role for argonauts in microRNA processing and posttranscriptional regulation of microRNA expression. *Cell* 131: 1097-1108, 2007.

[117] RC Friedman, KK Farh, CB Burge, and DP Bartel, Most mammalian mRNAs are conserved targets of microRNA. *Genome Res* 19: 92-105 (2009).

[118] Only a few ribozymes have been identified in humans and other mammals. In addition to catalyzing the splicing of other RNAs, some ribozymes cleave themselves, a process known as autocatalysis.

[119] LS Kristensen, TB Hanson et.al., Circular RNAs in cancer: opportunities and challenges in the field. *Oncogene* (2018) 37, 535-565.

[120] J Greene et.al., Circular RNAs: Biogenesis, Function and Role in Human Diseases. *Front Mol Bio*sci 2017 Jun 6;4:38. doi: 10.3389/fmolb.2017.00038. eCollection 2017.

[121] A more recent hypothesis stipulates that both RNA and DNA may have formed concurrently, prior to the formation of amino acids, in primordial times. Other lines of thinking suggest that the nucleic acids evolved concurrently with amino acids and proteins.

Chapter 6

[122] P Rous, A sarcoma of the foul transmissible by an agent separable from the tumor cells. *J Exp Med* 13: 397-411, 1911.

[123] There can be more extensive differences between the viral oncogene and its human homolog, such as the presence of sequences attached to the viral oncogene when it is transferred back into the host animal. In addition, amplification of the proto-oncogene, even in its non-mutated and non-oncogenic state (where the gene is expressed at normal levels), can result in an excess of the protein product of the gene. This can cause aberrant signaling that initiates oncogenesis.

[124] The antibody hypervariable regions and the variable regions of the TCR are examples of mammalian genes that have increased rates of mutation relative to most of the mammalian genes.

[125] Other commonly found oncogenes in human cancer include KRAS, MYC, BRAF and EGFR. KRAS and MYC, considered for decades as impossible to target- undruggable is the word that is used- are now under intensive investigation as potential targets due to advances in our understanding of how their protein products drive cancer progression.

[126] PK Vogt, Retroviral Oncogenes: A Historical Primer. *Nat Rev Cancer* 2012 September; 12(9): 639–648. doi:10.1038/nrc3320

[127] EBV is present in about 90% of the human population. A rare cancer called Burkitt's lymphoma, which is found in West African populations, is caused by EBV infection. Recently, intriguing data has emerged suggesting that EBV infection may act as the trigger for multiple sclerosis, resulting in an aberrant immune response to the virus that leads to
autoimmunity, a characteristic common to MS and other inflammatory diseases such as rheumatoid arthritis, ulcerative colitis, and psoriasis.

[128] PK Vogt, Retroviral Oncogenes: A Historical Primer. *Nat Rev Cancer*, 2012 September; 12(9): 639-648.

[129] P Rous, The Challenge to Man of the Neoplastic Cell. *Nobel Lecture*, December 13, 1966.

[130] ibid.

[131] ER Fearson, Human Cancer Syndromes: Clues to the Origin and Nature of Cancer. *Science* 278(5340):1043-1050 (1997). Another reference has it closer to 5%.

[132] CO Nordling, A new theory on the cancer-inducing mechanism. *Br. J. Cancer* 7(1): 68–72 (1953).

[133] ibid.

[134] ibid.

[135] ibid.

[136] A Sarasin, An overview of the mechanisms of mutagenesis and carcinogenesis. *Mutation Research* 544: 99-106 (2003).

[137] Recent measurements confirm the order of magnitude of this prediction, with about three mutations per round of DNA replication.

[138] C Tomasetti and B Vogelstein, Variation in cancer risk among tissues can be explained by the number of stem cell divisions. *Science* 347 (6217), 78-81 (2015).

[139] In the case of the lung, exposure to airborne particulates and toxins requires frequent replacement of damaged cells. Similarly, the colon is in contact with the food that we eat, with all the accompanying additives and toxins present therein.

[140] C Tomasetti and B Vogelstein, Variation in cancer risk among tissues can be explained by the number of stem cell divisions. *Science* 347 (6217), 78-81 (2015).

[141] C Tomasetti, L Li and B Vogelstein, Stem cell divisions, somatic mutations, cancer etiology, and cancer prevention. *Science* 355 (6331), 1330-4 (2017).

[142] ibid.

[143] MA Nowak and B Waclaw, Genes, environment and "bad luck." *Science* 355 (6331): 1266-7 (2017).

[144] Such variables include the probability of cancer initiation, probability of progression, rate of stem cell division (which likely varies over time), and

the number of stem cells.

[145] C Tomasetti and B Vogelstein, Variation in cancer risk among tissues can be explained by the number of stem cell divisions. *Science* 347 (6217), 78-81 (2015).

[146] C Tomasetti, L Li and B Vogelstein, Stem cell divisions, somatic mutations, cancer etiology, and cancer prevention. *Science* 355 (6331), 1330-4 (2017).

[147] LA Loeb, Human cancers express mutator phenotypes: origin, consequences and targeting. *Nature Rev Cancer* 11: 450-7, 2011.

[148] https://cancergenome.nih.gov

[149] LA Loeb, Human cancers express mutator phenotypes: origin, consequences and targeting. *Nature Rev Cancer* 11: 450-7, 2011.

[150] LA Loeb, CF Springgate, and N Battula, Errors in DNA replication as a basis of malignant change. *Cancer Res* 34, 2311-21, 1974.

[151] ABL1 is a proto-oncogene that can, like all proto-oncogenes, be converted into an oncogene when mutated in specific ways.

[152] The story of the discovery of the Philadelphia Chromosome is beautifully told in journalist Jessica Wapner's book *The Philadelphia Chromosome* (New York: The Experiment LLC, 2013).

[153] Point mutations in the ABL1 sequence of the hybrid kinase are often found in patients who develop resistance to Gleevec and other inhibitors of the kinase that have been developed since Gleevec's introduction in 1998.

[154] AJ Holland and DW Cleveland, Boveri revisited: chromosomal instability, aneuploidy and tumorigenesis. *Nat Rev Mol Cell Biol* 2009 July; 10(7): 478–487. doi:10.1038/nrm2718

[155] The relationship between gene copy number and the level of gene product is not always linear due to the complexities of the control of gene expression in living cells.

[156] EH Blackburn et.al., Human telomere biology: A contributory and interactive factor in aging, disease risks, and protection. *Science* 350 (6265): 1193-98 (2015).

[157] NW Kim et. al., Specific association of human telomerase activity with immortal cells and cancer. *Science* 266, 2011-15 (1994).

[158] SE Artandi, Telomeres and Telomerase in Cancer. *Carcinogenesis* vol. 31 no.1 pp.9–18, 2010.

Chapter 7

[159] AJ Holland and DW Cleveland, Boveri revisited: Chromosomal instability, aneuploidy and tumorigenesis. *Nat Rev Mol Cell Biol* 10(7): 478-87 (2009).

[160] Down's syndrome is an example of a human disease characterized by stable aneuploidy. The presence of an extra copy of chromosome 21 creates a stable condition called trisomy, in which there are 3 copies of a chromosome.

[161] HH Heng, SW Bremer, et. al. Chromosomal instability (CIN): what it is and why it is crucial to cancer evolution. *Cancer Metastasis Rev* 32: 325-40 (2013).

[162] TA Potapova, J Zhu, J and R Li, Aneuploidy and chromosomal instability: a viscous cycle driving cellular evolution and genome chaos. *Cancer Metastasis Rev* 32: 377-89 (2013).

[163] ibid.

[164] Vogelstein, B et al., Cancer Genome Landscapes. *Science* 339:1546-58 (2013).

[165] In phosphorylation, kinase enzymes add a phosphate group $[(PO_4)]^{3-}$ to one or more serines, threonines, and/or tryptophans on a target protein, which induces a conformational change in the protein that activates its function. For example, when the first kinase in a cascade phosphorylates the second kinase, the second protein undergoes a conformational change that results in its activation such that it can phosphorylate the third kinase in the cascade. Enzymes called phosphatases remove the phosphate groups from the kinase target proteins in a process called de-phosphorylation.

[166] Defects in this class of receptor are oncogenic in breast cancer (HER-2)

and colon cancer (EGFR), which are treated with anti-growth factor receptor antibodies (that is, antibodies to the growth factor receptors) such as Herceptin and Vectibix, respectively.

[167] D Hanahan and RA Weinberg, The Hallmarks of Cancer. *Cell* 100(1): 57-70 (2000).

[168] Such shortages might include components such as oxygen, glucose, and/or nucleotides for RNA and DNA synthesis.

[169] D Hanahan and RA Weinberg, The Hallmarks of Cancer. *Cell* 100(1): 57-70 (2000).

[170] D Hanahan and RA Weinberg, Hallmarks of Cancer: The Next Generation. *Cell* 144(5):646-74 (2011).

[171] M Enserink and E Pennisi, Nobel honors discoveries in how cells eat themselves. Science 354: 20 (2016).

[172] Y Ohsumi, Nobel Prize Biography. Nobelprize.org.

[173] ibid.

[174] E White and RS DiPaola, The Double-Edged Sword of Autophagy Modulation in Cancer. *Clin Cancer Research* 15: 5308-16 (2009).

Chapter 8

[175] ATP is a nucleotide comprised of the nucleoside adenosine—the nucleic acid base adenine linked to the sugar ribose—to which three phosphate groups are attached, all in a row. The hydrolysis (water-catalyzed bond breakage) of the bond between the second and third phosphorus atoms—the two most distant from the bond of the triphosphate to the ribose sugar—releases energy that is used for driving biological reactions.

[176] P Boyer, Energy, Life, and ATP. *Nobel Prize Lecture*, Dec. 10, 1997.

[177] S Vyas et al., Mitochondria and Cancer. *Cell* 166: 555-566 (2016).

[178] There are, in addition, RNAs in the cytoplasm of the egg that become part of the embryo, so this also adds a little more of "mom" to the mix.

[179] This is almost true. There are trace levels of other atoms in our macromolecules, such as the divalent cations (e.g., magnesium, manganese, zinc,

and copper atoms that bear a net charge of +2) that support the catalytic function of many enzymes. Other proteins also bind divalent cations; for example, calcium is essential for stabilizing Factor VIII, the protein that is deficient in the bleeding disorder Hemophilia A.

[180] Due to its metabolic importance, maintaining sufficient glutamine is required for a healthy cell culture.

[181] RA Brand, Biographical Sketch: Otto Heinrich Warburg, PhD, MD. *Clin Orthop Rel Res* (2010) 468:2831-2.

[182] O Warburg, F Wind, and E Negelein. The Metabolism of Tumors in the Body. *J Gen Physiol* 8:519-30 (1927).

[183] ibid.

[184] O Warburg, The prime cause and prevention of cancer- Part I. 1966. http://healingtools.tripod.com

[185] PS Ward and CB Thompson, *Metabolic Reprogramming: A Cancer Hallmark Even Warburg Did Not Anticipate. Cancer Cell* (2012) 21: 297-308.

[186] A single glucose molecule that undergoes glycolysis and fermentation to lactate generates two molecules of ATP, whereas that same glucose molecule can generate approximately 38 molecules of ATP if it is fully metabolized through the TCA cycle in the mitochondrion. These numbers are approximate, as the efficiency of biological systems is not 100%, and some energy can be lost along the way.

[187] A Luengo et al., Increased demand for NAD+ relative to ATP drives aerobic glycolysis. *Molecular Cell* 81: 1–17(2021).

[188] NN Pavlova et al., The Emerging Hallmarks of Cancer Metabolism (2016) *Cell Metabolism* 23, 27-47.

[189] S Peixoto da Silva et al., Cancer cachexia and its pathophysiology: links with sarcopenia, anorexia and asthenia. *J Cachexia Sarcopenia Muscle*. 2020 Jun; 11(3): 619–635.

[190] ibid.

[191] K Gupta et al., Oxygen regulates molecular mechanisms of cancer

progression and metastasis. *Cancer Metastasis Rev* 33: 183-215 (2014)

[192] ibid.

[193] The Nobel Prize in Physiology or Medicine was awarded in 2019 to the pioneers of the field of oxygen sensing, Gregg Semenza, Peter Radcliffe and William G. Kaelin.

[194] B Philip et al., HIF expression and the role of hypoxic environments within primary tumors as protective sites driving cancer stem cell renewal and metastatic progression. *Carcinogenesis* 34 (8): 1699-1707 (2013).

[195] LQ Chen and MD Pagel, Evaluating pH in the Extracellular Tumor Microenvironment Using CEST MRI and Other Imaging Methods. *Adv Radiol* 2015:206405

[196] ibid.

[197] D Ribatti, Judah Folkman, a pioneer in the study of angiogenesis. *Angiogenesis* 11:3-10 (2008).

[198] J Folkman, Tumor Angiogenesis: Therapeutic Implications. *N Engl J Med* 285:1182-1186 (1971).

[199] ibid.

[200] *The Oncologist* 2008; 13(2): 205-12.

[201] ibid.

Chapter 9

[202] Psychologists call this phenomenon "confirmation bias," characterized by the tendency to believe in ideas that support what we think we already know rather than accepting the possibility that ideas that contradict what we already know might be true.

[203] The story of the discovery of cells and the role of cells in biology and medicine is the subject of Siddhartha Mukherjee's *The Soul of the Cell: An Exploration of Medicine and the New Human (*New York: Charles Scribner's Sons, 2022).

[204] SY Tan and J Brown. *Rudolph Virchow (1821-1902): "Pope of pathology."* Singapore Med J 2006; 47(7):567-8.

[205] ibid.

[206] SI Grivennikov et al., *Immunity, Inflammation and Cancer*. Cell 140: 883-99 (2010).

[207] Such correlations between inflammatory conditions and cancer include viral hepatitis and liver cancer, chronic gastrointestinal disease and GI cancers (e.g., esophagus, stomach, and colon), and smoking and lung cancer.

[208] M Schafer and S Werner, Cancer as an overhealing wound: an old hypothesis revisited. *Nature Rev Mol Cell Bio* 9:628-38 (2008).

[209] C Rosales, Neutrophil: A Cell with Many Roles in Inflammation or Several Cell Types? *Front. Physiol* 9:113 (2018).

[210] L Erpenbeck and MP Schon, Neutrophil extracellular traps: protagonists of cancer progression? *Oncogene* 36 (18): 2483-90 (2017).

[211] SA Patel et al., Inflammatory Mediators: Parallels between cancer biology and stem cell therapy. *J Inflammation Res* 2:13-19 (2009).

[212] D Hanahan and RA Weinberg, Hallmarks of cancer: the next generation. *Cell* 2011 144(5): 646-74 (2011).

[213] JB Gurdon and JA Byrne, The first half-century of nuclear transplantation. *Proc Nat Acad Sci USA* 100 (14):8048-52 (2003).

[214] R Briggs and TJ King, *Transplantation of living nuclei from blastula cells into enucleated frog eggs*. Proc Natl Acad Sci USA 38:455-63 (1952).

[215] Xenopus is a widely used model due to its short lifecycle and responsiveness to human hormones in the laboratory.

[216] We now know that the very early embryonic cells are totipotent cells, and the level of pluripotency of the subsequent generations of cells declines rapidly. This process is evidently reversible.

[217] Those of us who were around back then remember that the birth of a "test tube" sheep named Dolly created a storm of conversation in the press, including dire warnings of the imminent derangement of the technology to create designer babies.

[218] My doctoral dissertation described the purification and properties of a histone acetyltransferase from hog livers, which I obtained fresh from a

local sausage factory in Durham, NC. The livers were taken to the lab on dry ice and homogenized in a large high-speed blender to create a brown sludge that I called the "hog liver milkshake." One day I learned the critical importance of making sure that all four metal clamps on the top of the blender were securely fastened. The cold room in which I worked to protect the biochemical structure of the fragile enzyme from heat was a nauseating landscape, the counters, floor and ceiling painted with splatters of blood and gristle. It was a scene reminiscent of a zombie apocalypse. I never repeated that mistake.

[219] JD Minna and JE Johnson, Opening a Chromatin Gate to Metastasis. *Cell* 166: 275-6 (2016).

[220] ibid.

[221] ND Marjanovic, RA Weinberg and CL Chaffer, Poised with purpose: Cell plasticity enhances tumorigenicity. *Cell Cycle* 12: 2713-4 (2013).

Chapter 10

[222] Y Li and J Laterra, Cancer Stem Cells: Distinct Entities or Dynamically Regulated Phenotypes. *Cancer Res 72(3): 576-80 (2012).*

[223] JC Recamier. *Recherches sur le traitement du cancer sur la compression methodique simple ou combinee et sur l'histoire gererale de la meme maladie*, 2nd Edition. Paris: Chez Gabon, 1829.

[224] ibid.

[225] JE Talmadge and IJ Fidler, AACR Centennial Series: The Biology of Metastasis: Historical Perspective. *Cancer Res* 70(14): 5649-69 (2010).

[226] ibid.

[227] ibid.

[228] L Mathot and J Stenninger, Behavior of seeds and soil in the mechanism of metastasis: A deeper understanding. *Cancer Sci* 103(4):626-31 (2012).

[229] MA Nieto et al., EMT: 2016. *Cell* 166: 21-45 (2016).

[230] P Friedl, Y Hegerfeldt and M Tusch, Collective cell migration in morphogenesis and cancer. *Int J Dev Biol* 48: 441-9 (2004).

[231] JM Westcott et al., An epigenetically distinct breast cancer cell subpopulation promotes collective invasion. *J Clin Invest* 2015;125(5):1927-1943.

[232] A matrix metalloproteinase is an enzyme (a *proteinase*, aka *protease*) that degrades proteins. The enzyme activity is dependent on the presence of a specific metal (*metallo-*), and the enzyme degrades the protein components of the extracellular *matrix*.

[233] The tendency of most solid tumors to metastasize to a limited set of organ systems may depend, at least in part, on the ability of exosomes to target specific cell types with their molecular payloads.

[234] M Tkach and C Thery, *Communication by Extracellular Vesicles: Where We Are and Where We Need to Go. Cell* 164: 1226-32 (2016).

[235] ibid.

[236] D Hanahan and RA Weinberg. The Hallmarks of Cancer. *Cell* 100(1): 57-70 (2000).

[237] D Hanahan and RA Weinberg. Hallmarks of Cancer: The Next Generation. *Cell* 146(5): 646-74 (2011).

Chapter 11

[238] Building on this principle, a new approach to chemotherapy that attempts to "push cancer cells over the edge" by using their vulnerabilities against them will be described in a later chapter.

[239] When we use the word "cure" with reference to cancer, we need to consider the impossibility (at least with current or foreseeable technology) to definitively prove the absence of all cancer cells from the body and that there is zero risk of their future return. The absence of detectable cancer cells in a patient is therefore characterized as a remission, regardless of the duration of the "cancer-free" state.

[240] RJ Gillies, D Verduzco and RA Gatenby, Evolutionary dynamics of carcinogenesis and why targeted therapy does not work. *Nature Rev Cancer* 12: 487-93 (2012).

[241] ibid.

[242] ibid.

[243] Y Sun and PS Nelson, Molecular Pathways: Involving Microenvironment Damage Responses in Cancer Therapy Resistance. *Clin Cancer Res* 18(15): 4019-25 (2012).

[244] ibid.

[245] While the plasticity and unpredictability of cancer cells are at the root of the extraordinary challenges in treating human cancer, these properties also make cancer, from a scientific standpoint, the most fascinating of all human ailments.

[246] Membranes are comprised of both uncharged (hydrophobic, or "water-hating") as well as charged (hydrophilic, or "water-loving") molecules, thereby limiting the number of substances that can freely pass through without interacting with the membrane components. This significantly limits the permeability of the membrane to substances in the interstitial space outside the cell, including toxins.

[247] MM Gottesman, Mechanisms of Cancer Drug Resistance. *Annu Rev Med* 53: 615-27 (2003).

[248] ibid. Chemotherapeutic agents that bind to P-glycoprotein include commonly used drugs such as taxol, doxorubicin, vinblastine, and vincristine.

[249] LJ Goldstein et al., Expression of Multidrug Resistance Gene in Human Cancers. *Journal of the National Cancer Institute* 81(2): 116-24 (1989).

[250] MM Gottesman, Mechanisms of Cancer Drug Resistance. *Annu Rev Med* 53: 615-27 (2003).

[251] ibid.

[252] Y Sun and PS Nelson, Molecular Pathways: Involving Microenvironment Damage Responses in Cancer Therapy Resistance. *Clin Cancer Res* 18(15): 4019-25 (2012).

[253] A microenvironment that is resistant to therapeutic intervention is called a chemoresistant niche.

[254] Replicative exhaustion may or may not involve shortened telomeres as a root cause; senescence can occur due to environmental stresses even

without the cell reaching the "Hayflick limit."

[255] Y Sun and PS Nelson, Molecular Pathways: Involving Microenvironment Damage Responses in Cancer Therapy Resistance. *Clin Cancer Res* 18(15): 4019-25 (2012).

[256] ibid.

[257] M Maugeri-Sacca, P Vigneri and R De Maria, *Cancer* Stem Cells and Chemosensitivity. Clin. Cancer Res. 17(15): 4942-6 (2011). M Dean, T Fojo and S Bates, *Tumor* Stem Cells and Drug Resistance. *Nature Rev Cancer* 5:275-84 (2005).

Chapter 12

[258] LF Vernon, William Bradley Coley, MD, and the phenomenon of spontaneous regression. *ImmunoTargets and Therapy* 7: 29-34 (2018).

[259] Laziosi was canonized as Saint Peregrine in 1726 and is recognized by the Roman Catholic Church as the patron saint of cancer patients.

[260] P Kucerova and M Cervinkova, Spontaneous regression of tumor and the role of microbial infection- possibilities of cancer treatment. Anti-Cancer Drugs 27(4): 269-77 (2016).

[261] EF McCarthy, The toxins of William B. Coley and the Treatment of Bone and Soft-Tissue Sarcomas. *The Iowa Orthopedic Journal* 26:154-8. (2006).

[262] LF Vernon, William Bradley Coley, MD, and the phenomenon of spontaneous regression. *ImmunoTargets and Therapy* 7: 29-34 (2018).

[263] ibid.

[264] These successes included the case of an Italian immigrant named Emile Zola (not the French novelist), whose recurrent sarcoma of the neck and tonsil left him only weeks from death when he began to receive injections of Coley's toxin. Following treatment, Mr. Zola's sarcoma remained in remission for eight years.

[265] EF McCarthy, The toxins of William B. Coley and the Treatment of Bone and Soft-Tissue Sarcomas. *The Iowa Orthopedic Journal* 26:154-8

2006.

[266] SA Hoption Cann, JP van Netten, and C van Netten, Dr. William Coley and tumor regression: a place in history or in the future. *Postgrad Med J* 79:672-80 (2003).

[267] GS Kienle, Fever in Cancer Treatment: Coley's Therapy and Epidemiological Observations. *Global Advances in Health and Medicine* 1(1): 92-99 (2012).

[268] ibid.

[269] ibid.

[270] ibid.

[271] ibid.

[272] RD Schreiber, LJ Old, and MJ Smyth, Cancer Immunoediting: Integrating Immunity's Roles in Cancer Suppression and Promotion. *Science* 331: 1565-70 (2011).

[273] GA Currie, Eighty Years of Immunotherapy: A Review of Immunological Methods Used for the Treatment of Human Cancer. *Br J Cancer* 26: 141-51 (1972).

[274] RR Smith et al., Studies on the use of viruses in the treatment of carcinoma of the cervix. *Cancer* 9: 1211-18 (1956).

[275] R Nuwer, Viruses Recruited as Killers of Tumors. *The New York Times,* March 18, 2012.

[276] To the persistent scientist, an experimental dead-end does not engender surrender. The lessons learned from failure might (if the stars are properly aligned) lead to a new understanding of the problem. By finding out what does not work (and more critically, why it didn't work), it is sometimes possible to open the door to more promising approaches.

[277] SA Rosenberg and NP Restifo, Adoptive cell transfer as personalized immunotherapy for cancer. *Science* 348: 62-8 (2015).

[278] J Cavallo, Steven A. Rosenberg, MD, PhD, Works to Unmask Cancer's Achilles Heel. *The ASCO Post,* November 25, 2018.

[279] E Petrus, An Immunotherapy Pioneer Tells All.

https://irp.nih.gov/catalyst/26/1/steven-a-rosenberg-md-phd

[280] SA Rosenberg, NP Restifo, JC Yang, RA Morgan, and ME Dudley, Adoptive cell transfer: a clinical path to effective cancer immunotherapy. *Nat Rev Cancer* 8(4): 299-308 (2008).

[281] EJ Delorme and P Alexander, Treatment of primary fibrosarcoma in the rat with immune lymphocytes. *Lancet* 2: 117-120 (1964); A Fefer, Immunotherapy and chemotherapy of Moloney sarcoma virus-induced tumors in mice. *Cancer Res* 29: 2177-83 (1969).

[282] ibid.

[283] DA Morgan, FW Ruscetti, and R Gallo, Selective *in vitro* Growth of T Lymphocytes from Human Bone Marrows. *Science* 193 (4257): 1007-8 (1976).

[284] SA Rosenberg and NP Restifo, Adoptive cell transfer as personalized immunotherapy for human cancer. *Science* 348 (6230):62-8 (2015).

[285] E Petrus, *An Immunotherapy Pioneer Tells All*. https://Irp.nih.gov/catalyst/26/1/steven-a-rosenberg-md-phd

[286] J Cavallo, Steven A. Rosenberg, MD, PhD, Works to Unmask Cancer's Achilles Heel. *The ASCO Post*, November 25, 2018.

[287] ibid.

[288] OJ Finn, Human Tumor Antigens Yesterday, Today and Tomorrow. *Cancer Immunol Res* 5:347-54 (2017).

[289] ibid.

[290] EL Reihherz, Revisiting the discovery of the TCR complex and its co-receptors. *Frontiers in Immunology* 5: 1-4 (2014).

[291] ibid.

[292] ibid.

[293] EL Reinherz, Revisiting the discovery of the TCR complex and its co-receptors. *Frontiers in Immunology* 5:1-4 (2014).

[294] ibid.

[295] ibid.

[296] ibid.

[297] Ibid.

[298] SC Meuer et al., Clonotypic Structures Involved in Antigen-Specific Human T Cell Function. *J Exp. Med* 157: 705-719 (1983); SC Meuer et al., Identification of the Receptor for Antigen and Major Histocompatibility Complex on Human Inducer T Lymphocytes. *Science* 222: 1239-41 (1983)

[299] As there are many alleles for each MHC gene, the likelihood of an exact match for all six MHC genes (three each for class I and class II) is low. The likelihood decreases in more heterogeneous populations derived from different ethnic and geographical locations, as MHC expression patterns are more similar amongst individuals in more homogenous populations.

[300] This estimate was made two decades ago by my dear colleague and friend at Amgen, Dr. Izydor Apostol.

[301] TN Schumacher and RD Schreiber, Neoantigens in cancer immunotherapy. *Science* 348(6230): 69-74 (2015).

[302] ibid.

[303] A mutation that does not alter the amino acid sequence in a protein is called a silent mutation. If the mutation results in a change in the amino acid sequence, that is called a missense mutation, and if the change introduces a stop codon that terminates the protein chain prematurely, that is called a nonsense mutation.

[304] ibid.

Chapter 13

[305] The desensitization response involves multiple mechanisms, including lower levels of expression of the receptor and internalization of existing surface receptors followed by recycling of their amino acid components for the biosynthesis of new proteins.

[306] In addition, tumor infiltrates often include other immunosuppressive cell types, such as mesenchymal-derived stromal cells (MDSC), tumor-associated macrophages (TAMs), and cancer-associated fibroblasts (CAFs) that are co-opted by the tumor microenvironment to secrete

immunosuppressive proteins.

[307] The screening procedure is performed in plastic immunoassay plates that can be coated with an extract of intact tumor cells containing the surface proteins of the tumor.

[308] SA Rosenberg and NP Restifo, Adoptive cell transfer as personalized immunotherapy for human cancer. *Science* 348(6230): 62-8 (2015).

[309] ibid.

[310] SA Rosenberg et al., Durable Complete Responses in Heavily Pretreated Patients with Metastatic Melanoma Using T Cell Transfer Immunotherapy. *Clin Cancer Res* 2011 July 1; 17(13): 4550–4557.

[311] SA Rosenberg et.al., Adoptive cell transfer: a clinical path to effective cancer immunotherapy. *Nat Rev Cancer* 8(4): 299-308 (2008).

[312] JH van den Berg et al., Tumor infiltrating lymphocytes (TIL) in metastatic melanoma: boosting of neo-antigen specific T cell reactivity and long term follow-up. *J Immunotherapy of Cancer* 8:e000848 (2020).

[313] It is easy to confuse the terms "avidity" and "affinity." The former refers to the strength of the interaction between an antigen and antibody. In contrast, affinity is based on the rates at which the antigen binds to and dissociates from the antibody (the ratio of the rate of binding and rate of dissociation is called the equilibrium constant).

[314] It is now possible to remove the genes for the native TCR from the transduced cell so that it only expresses the engineered TCR.

[315] Medical practitioners called hematologists treat blood disorders (also known as hematological disorders) such as leukemia, anemia, sickle cell disease and hemophilia.

[316] ibid.

[317] 'We have to cure' cancer, says CAR-T pioneer Carl H. June, MD. *HemOnc Today,* April 18, 2019.

[318] Therapeutic monoclonal antibodies are sometimes designed to both bind to a target and elicit effector functions, such as cell killing mediated by T cells. The leukemia drug Rituxan is an example. Cell killing by antibodies

is highly localized and of short duration, whereas stimulation of a cytotoxic T cell response (the goal of ACT) can provide durable, long-term, and systemic immune surveillance.

[319] In the case of direct presentation of antigens to T cells by tumor cells, proteins on the surface of the tumor cell also interact with proteins on the surface of the T cells.

[320] J Couzin-Frankel, The Dizzying Journey to a New Cancer Arsenal. *Science* 340: 1514-18 (2013).

[321] S Guedan et al., Emerging Cellular Therapies for Cancer. *Annu Rev Immunol* 37: 145-71 (2019).

[322] CRS has also been seen in COVID-19 patients due to over-stimulation of the immune system by the viral infection.

[323] J Couzin-Frankel, The Dizzying Journey to a New Cancer Arsenal. *Science* 340: 1514-18 (2013).

[324] CH June et al., CAR-T cell immunotherapy for human cancer. *Science* 359:1361-5 (2018).

[325] RC Sterner and RM Sterner, CAR-T cell therapy: current limitations and potential strategies. *Blood Cancer Journal* 11:69 (2021).

[326] In these patients, B cell function can be depleted for months or even years, depending on the duration of the CAR-T cells in the body. Over time, B cell function returns as the hematopoietic stem cells reconstitute the populations of cells damaged by the treatment. The same is true for patients receiving the monoclonal antibody Rituxan, which targets CD20 on B cells. Rituxan is used to treat non-Hodgkin's lymphoma and other B cell malignancies.

[327] The engineered T cell technology was also used by June and his team as a treatment for HIV, resulting in the first demonstration of the elimination of HIV from patients.

Chapter 14

[328] DR Leach et al., Enhancement of Antitumor Immunity by CTLA-4

Blockade. *Science* 271 (5256): 1734-36 (1996).

[329] ibid.

[330] ibid.

[331] "The Texas T cell Mechanic," *Cancer Research Institute*. www.cancer-research.org/immunotherapy/stories/scientists/james-p-Allison-phd

[332] The tumors resulted from injection of colon carcinoma cells in the flanks of the mice (below the hips).

[333] The discovery of the underlying scientific mechanisms at play in the immune checkpoints illuminated by Dr. Allison and Dr. Honjo resulted in the presentation of the 2016 Nobel Prize in Physiology or Medicine to the two scientists who characterized these two key immunomodulatory mechanisms.

[334] Cytokine storms are responsible for patient outcomes (often dire) in septic shock that result from an out-of-control immune response to infection. As an example, they significantly contribute to mortality in severe infection with COVID-19 and pandemic flu.

[335] RM Condry et al., Talimogene laherparepvec: First in class oncolytic virotherapy. *Human Vaccines & Immunotherapeutics* 2018, Vol. 14, No. 4, 839–846.

[336] HL Kaufmann et al., Oncolytic viruses: a new class of immunotherapy drugs. *Nature Reviews Drug Discovery* 14: 642-62 (2015).

[337] RM Condry et al., Talimogene laherparepvec: First in class oncolytic virotherapy. *Human Vaccines & Immunotherapeutics* 2018, Vol. 14, No. 4, 839–846.

[338] Ibid.

[339] H Fukuhara et al., Oncolytic virus therapy: A new era of cancer treatment at dawn. *Cancer Sci* October 2016; 107(10): 1373–1379.

[340] ibid.

[341] RM Condry et al., Talimogene laherparepvec: First in class oncolytic virotherapy. *Human Vaccines & Immunotherapeutics* 2018, Vol. 14, No. 4, 839–846.

[342] O Hemminki et al. Oncolytic viruses for cancer immunotherapy. *Journal of Hematology & Oncology* (2020) 13: 84.

[343] The original polio vaccine, invented by Dr. Jonas Salk in the 1950s, was administered by injection.

[344] While the world today is filled with naysayers and skeptics in such matters, I doubt there were significant numbers of parents protesting the arrival of a wondrous new medicine that protected children (and adults) from a devastating illness. Images of children in iron lungs on television and in magazines provided a powerful incentive for widespread acceptance of the polio vaccine. Ironically, today's vaccines are orders of magnitude safer and far more thoroughly tested than those available back then.

[345] A. Lopes et al., Cancer DNA Vaccines: current preclinical developments and future perspectives. *J Exp Clinical Cancer Res* 38:146 (2019).

[346] ibid.

[347] K Vermaelen, Vaccine strategies to improve anti-cancer immune responses. *Frontiers in Immunology* 10 (Article 8) (2019).

[348] Experts Perspective: Dr Ira Pastan, *Chinese Antibody Society*, March 12, 2021.

[349] RJ Kreitman et al., Phase I trial of anti-CD22 recombinant immunotoxin Moxetumomab Pasudotox (CAT-8015 or HA22) in patients with hairy cell leukemia. *J Clinical Oncology* 30(15): 1822-28 (2012).

[350] C Worthington, FDA-Approved Drug is a Home Run Several Decades in the Making. https://ncifrederick.cancer.gov/about/theposter/content/fda-approved-drug-home-run-several-decades-making, February 20, 2019.

[351] This immunotoxin construct is comprised of the same type of single-chain variable fragment (scFv) as used in CAR-T constructs. As the *Pseudomonas* toxin has both a cell docking and cell killing domain, and the cell docking function of the toxin is not needed due to the presence of the antibody fragment, only the cell killing domain of the toxin is linked to the scFv antibody fragment. Since smaller proteins can more easily penetrate

tumors than larger ones, the goal is to use only the pieces of the antibody and toxin needed for targeting and killing tumor cells. While removing the Fc domain of the antibody significantly shortens its half-life, this can be advantageous with such a toxic drug, as it can clear from the patient's body in days instead of weeks following discontinuation due to unacceptable toxicity.

[352] A Thomas et al., Antibody-drug conjugates for cancer therapy. *Lancet Oncology* 17(6): e254–e262 (2016).

[353] In some ADCs, the amino acid sequence that bridges the antibody to the toxin is designed such that it can be cleaved by proteases in the lysosomes. When the bridge sequence is cleaved, the toxin is liberated inside the cell.

[354] These immunological effects can usually be managed effectively with immunosuppressive medications, which do, however, make the patient more vulnerable to infection by pathogenic organisms.

[355] S Ponziani et al., Antibody-Drug Conjugates: The New Frontier of Chemotherapy. *Int J Mol Sciences* 21: 5510 (2020).

Chapter 15

[356] 'We have to cure' cancer, says CAR-T pioneer Carl H. June, MD. *HemOnc Today*, April 18, 2019. https://www.healio.com/news/hematology-oncology/20190418/we-have-to-cure-cancer-says-car-t-pioneer-carl-h-june-md

[357] ibid.

[358] The current generation of CAR-T therapeutics cost in the range of about $375,000 to $475,000 per patient. The cost is likely to come down significantly if the cost of manufacturing operations for these patient-specific therapies can be reduced through technological innovation and/or if the allogeneic, off-the-shelf approach is workable, at least for some patients.

[359] JH Park et al., Cancer Metabolism: Phenotype, Signaling and Therapeutic Targets. *Cells* 2020, 9, 2308; doi:10.3390/cells9102308

[360] ibid.

[361] P Farhadi et al., The emerging role of targeting cancer metabolism for cancer therapy. *Tumor Biology* October 2020: 11-18 (2020).
[362] ibid.
[363] A series of scientific meetings, including a historic conference at Asilomar, CA, led to a consensus prohibiting the release of organisms competent for survival outside of the laboratory. Other provisions were agreed to that were designed to limit the risk of the release of genetically modified organisms into the environment.
[364] For an excellent read on the controversies raised by the development of gene editing, read Walter Isaacson's *The Code Breaker*.
[365] S Guedan et al., Emerging Cellular Therapies for Cancer. *Ann Rev Immunol* 37: 145-71 (2019).
[366] In April of 2023, a team of scientists at Johns Hopkins University reported results from a study in mice of a new method of delivering cancer drugs directly into tumors. By coating the tumor bed in a gel material (called a hydrogel) that contains the chemotherapy drug paclitaxel and an antibody to the CD47 protein, they obtained tumor eradication in every mouse tested. This intriguing finding is both exciting and a long way from human trials. See E. Cara, Experimental Gel Killed 100% of Brain Tumors in Mice. *Gizmodo (Apple News)*, April 24, 2023.
[367] F Cheng et.al., Circulating tumor DNA: a promising biomarker in the liquid biopsy of cancer. *Oncotarget* 7(30): 48832-41 (2016); A Campos-Carrillo, Circulating tumor DNA as an early cancer detection tool. *Pharmacol Ther* 107458 (2019).
[368] The Cologuard screening test for colon cancer is an example of the use of circulating tumor DNA for early diagnosis.
[369] A recent development in early cancer detection that may add another potent weapon to the liquid biopsy arsenal is based on the detection of antibodies in the blood that are highly correlated with certain types of tumors.
[370] A recent publication from the UK identified "mutational signatures" that appear to be characteristic of cancers caused by the same process—for

example, lung cancers caused by cigarette smoking and melanomas caused by UV damage.

[371] The CRISPR technology is being used to create highly sensitive diagnostic techniques for cancer and other diseases.

[372] S Guedan et al., Emerging Cellular Therapies for Cancer. *Annu Rev Immunol* 37:14571 (2019).

[373] ibid.

[374] I Williams, Breakthrough cancer treatment nearly wipes out tumor in days. https://talker.news, March 13, 2024.

[375] ibid.

[376] SF Ngiow and A Young, Re-education of the tumor microenvironment with targeted therapies and immunotherapies. *Front Immunol* 11: 1633 (2020).

[377] The same delivery technology used in the COVID mRNA vaccines could be used to deliver medical payloads to cancer cells. This technology, called lipid nanoparticles, is highly adaptable for use in multiple applications. For example, the nanoparticles can be studded with antibodies to target them to specific cell types, including cancer cells.

[378] Recent evidence shows that bacteria and other microbes also reside in our organs.

[379] A trillion is 10^{12}; a quadrillion is 10^{15}. If you'd really like to nerd out with me, 10^{18} is a quintillion. We could go on all day.

[380] D Lambert, Microbiome Research: Introducing the Archaeome. medicalnewstoday.com, January 24, 2012. https://www.medicalnewstoday.com/articles/microbiome-research-introducing-the-archaeome

[381] GD Sepich-Poore et al., The microbiome and human cancer. *Science* 371: eabc4552 (2021).

[382] S Rezasoltani et al., Modulatory effects of gut microbiome in cancer immunotherapy: A novel paradigm for blockade of immune checkpoint inhibitors. *Cancer Medicine* 10:1141-54 (2021).

[383] While colon cancer is the most prominent example in this context, recent

evidence indicates that microbes are present throughout the body, and dysbiosis may play a role in other epithelial cancers, such as carcinomas of the lung, pancreas, bladder, skin, kidney, and liver. Dysbiosis is also linked to many of the common inflammatory diseases, such as diabetes, obesity, rheumatoid arthritis, and irritable bowel syndrome.

[384] An example of an oncomicrobe is a bacterium called *Heliobacter pylori*, which has been directly linked to ulcers and stomach cancers.

[385] GD Sepich-Poore et al., The microbiome and human cancer. *Science* 371: eabc4552 (2021).

[386] ibid.

[387] These observations seem to fly in the face of the long-held belief that our organs are sterile environments free of microbial species.

[388] A research group at Harvard reported that when microbiome samples from immunotherapy responders and non-responders were implanted in tumor-bearing mice, the mice responded similarly to the human patient from whom the microbiome sample originated. The data pointed to the role of an immune checkpoint protein (and molecular cousin of PD-L1) called PD-L2 in the mechanism by which microbial metabolites impact the immune response to cancer. See To boost cancer immunotherapy's fighting power, look to the gut. *Harvard Medical School News Release*, May 3, 2023.

[389] Bacterial species can also assist the tumor in invading nearby tissues and/or interfere with chemotherapeutic agents. A species called *Fusobacterium nucleatum* that is normally found in the mouth (and is a major cause of periodontal disease) is a dominant species in many cases of colorectal cancer, with a demonstrated correlation between higher tumor loads of this bacterium and poor outcomes.

[390] ibid.

[391] ibid.

[392] Probiotics are the microorganisms themselves, while prebiotics are the substances that feed microbes and promote a healthy balance of microbial

species. Examples of probiotics include yogurt, fiber, and fermented foods like sauerkraut. Since the microbiome is a complicated mixture of species, probiotics should contain a range of organisms, as the administration of one or a limited number of types can promote their over-growth relative to the rest of the microbiome.

[393] The tragedy here is that a for-profit health care system is inequitable, such that quality health care is less available for the less fortunate.

Index

Numbers in italics refer to the illustrations.

A

acidosis, 152, 183, 188
adaptive immunity, 7-9, 12, 214, 216, 219-20, 224, 230
adenosine triphosphate (ATP), 141, *142*, 143–44, 148, 152, 192
adoptive cell transfer (ACT), 212, 223
Allison, James, 241–43
aneuploidy, 122–24, 129–31
angiogenesis, 154, 156, 182, 194, 268
antibodies
   antibody diversity, 10, 12, 15–16, 18
   antibody-drug conjugates (ADCs), 256–59
   antibody-producing cells, 9–12, 16–17
   disulfide bonds in, 13
   domains of, 15
   hinge of, 13
   monoclonal, 18, 179, 188
   polyclonal, 18
   structure of, 13, *14*, 15

anti-codon, 76
antigen-binding domains, 15, 229, 232-33
antigen presentation, 220, 246
   description of, 22
   effect of tumor cells on, 230
   HLA restriction of, 228, 231-32
   TCR binding during, 218, 241
antigen-presenting cells (APCs), 22, 221, 241-42
antigen stimulation, 8, 12
anti-interleukin-6, 264
anti-toxins, 4, 7, 9–11, 19
apoptosis
   cellular quiescence and, 55, 62
   definition of, 51
   p53 protein and, 126
   signaling of, 95, 117
   tumor suppression by, 137
autophagy, 138, 190
Avastin, 20, 156

B

base-pairing rules, 43-44, 54,

71, 76, 85
B cells, 10–12, 238, 243, 252
   activation of, *11*, 12
   memory B cells, *11*, 12
   naïve, 12
   See also antibodies and hybridoma
bifunctional antibodies, 252–53, 259
biopharmaceuticals, vii, 19, 88, 229
biotechnology, 88, 268
Bishop, Michael J., 105
Boveri, Theodore, 122
Boyer, Paul, 143
Brenner, Sydney, 73, 94–95, 139
Burnet, Frank Macfarlane, 10–12

C

cachexia, 149–50
cancer causation, 56, 113–15
cancer driver genes, 131–32
Cancer Genome Atlas, 117-18
cancer immunotherapy, 24, 213, 262, 272
   early attempts at 206
   neoantigens and, 222
   oncolytic viruses and, 247
   strategy of, 211
   vaccines and, 250–51
cancer incidence, 20, 111, 113-14

cancer prevention, 278-80
cancer progression. *See* tumor progression
cancer stem cells, 169–70, 261
   characteristics of, 196
   quiescent, 146, 187
cancer therapeutics, 19, 261
   delivery of, 253
   design of, 260
   efficacy of, 190, 230
   immunotherapeutics, 163, 234, 259, 264–65
   T cell therapeutics, 231-32
   protein therapeutics, 19, 198, 261
   resistance to, 189, 193, 260, 274–75
cancer treatment paradigm, 261
cancer types, 114, 117, 132, 208, 248
carcinogenesis, 59
   definition of, 52
   epigenetics and, 69
   genetic damage and, 116, 130
   inflammation and, 107, 160, 162
   lifestyle and, 278
   mechanisms of, 31
   oncogenes and, 107–8
   smoking and, 21
   telomeres and, 126
carcinomas, 175-76, 202, 205
   mammary, 169

Index

caretaker functions, 51, 55, 59, 118-19, 132, 187
   damage to, 117, 195
CAR-T cell. *See* chimeric antigen receptor T cell
Carter, Jimmy, 244
cascade
   apoptotic, 137
   coagulation, 161
CD3, 253
CD4, 216, 234
CD4+ helper cells, 216, 221, 224, 234, 274
CD8+ cells, 211, 216, 220, 224, 274
CD28, *234*, 236, 241-42
cell clones, 168
   invasive cancer, 262
   virulent cancer, 195
cell cycle, *53*, 54-55
   arrest of, 62, 126, 194
   cellular quality control of, 112, 116
   phases of, 52
   regulation of, 59, 62, 107, 109, 116, 135-36
cell fusion, 17, 58
cells
   allogeneic, 270
   aneuploid, 130-31
      See also aneuploidy
   daughter, 53-54, 131
   differentiated, 29, 32, 34, 163-65
   diploid, 123
   epithelial, 173-75, 190
   infected, 22, 62, 245
   lymphoid, 208, 226
   myeloid, 33, 226
   myeloma, 17
   quiescent, 185, 187, 261
cellular adhesion, 195
cellular differentiation
   definition of, 32-35
   process of, 163
   reversibility of, 169
cellular plasticity, 164, 182
   of cancer cells, 148, 261, 265
   definition of, 35
   as driver of cancer progression, 179, 262, 266
   of stem cells, 164, 168
cellular proliferation, 121, 132, 136, 139, 190, 194, 245
   control of, 107-9
   rapid, 108, 135
   uncontrolled, 59, 135
cellular senescence, 62, 113, 126, 195
Central Dogma of Molecular Biology, *70*, 71-75, 78
Charpentier, Emmanuelle, 267
chemical bonds, 41, 143-44
chemokines, 177
chemotherapeutic agents, 139, 187, 193, 212, 217, 256,

263
chemotherapy, 185–87, 203, 214, 226, 253, 256
  cancer cell response to, 139, 193, 217
  cellular damage by, 194, 261
  complications of, 190
  systemic, 263
  traditional, 194
chimeric antigen receptor T cell (CAR-T), *235*, 236-239, 264, 273-74
  allogeneic, 270
  engineering of, 269-70
  solid tumor treatment, 238
  target loss, 247
chromatin, 65, *66*, 67–69, 166
  architecture of, 92, 166–67, 182
  complexity of, 166
chromosomal instability, 126, 131
chromosomal shattering, 129
chromothripsis. *See* chromosomal shattering
circulating tumor DNA (ctDNA), 271
clinical trials, 19, 229, 239, 248, 253, 257, 272
Claude, Albert, 73
clonal selection theory, 12
codons, 76–78
cognate antigen, 232-33

Coley, William Bradley, 200–204
Coley's Toxin, 202, 250
Collins, Francis, 83
colon cancer, 20, 63, 113, 220, 224, 272
co-stimulatory proteins, *234*, 236
covert cancers, 23
COVID-19 pandemic, 140, 250
Crick, Francis, 78
  DNA discovery by, 36, 47
  mRNA discovery by, 94
  See also Central Dogma of Molecular Biology
CRISPR-Cas 9, 267
CTLA-4, *234*, 241–43, 265
cytokine release syndrome (CRS), 237, 264
cytokines, 97, 189, 195, 212, 224, 245, 249, 252, 269
  anti-inflammatory, 245
  immunosuppressive, 251
  interleukin-2, 211-12, 225
  interleukin-10, 251
  pro-inflammatory, 149, 162, 225, 245, 248, 277
cytokinesis, 54, 131
cytoplasm, 71–73, 98, 193, 247, 275

D

Dalton, John, 40, 62

Index

dendritic cells, 22, 246, 251, 278
diagnosis of cancer, 18, 57-58, 88, 185, 271
diet, healthy, 278-79
differentiated cells, 30, 34, 163, 165, 169-70, 196
disseminated tumor cells, 173, 177-78
DNA
  base pairs, 65, 67, 91
  damaged, 20, 51, 62, 116, 187
  double-stranded, 97
  human, 82, 106, 267
  single-stranded complementary, 86
DNA airport, 50, 63, 92, 166-67
DNA-binding proteins, 63, 92, 174
DNA damage, 51, 62, 186-88, 194
DNA Damage Response (DDR), 51, 116, 125, 194
DNA polymerases, 54, 85-86, 125
DNA repair, 62, 187, 194
DNA replication and cell division, 49, 52-53, 112, 116, 125, 130
DNA sequencing, 82-83, 91, 227-28, 260, 271
DNA structure, 43, 70-71, 76
  exons, 78
  introns, 78, 93

DNA synthesis, 54, 125
DNA viruses, 97
double-stranded RNA, 96-98
Doudna, Jennifer, 267
driver mutations, 116-18, 132, 261, 272
drug resistance, 260, 275
drugs
  biopharmaceuticals, xiv-xvi, 19, 88, 229, 253
  oncology, 264
durable remissions, 22, 266
durable response rate, 249
dysbiosis, 276
dysregulation
  immune, 107
  metabolic, 149, 183

E

effector functions of IgG, 15, 232
Ehrlich, Paul, 3, 6, 10, 18-19, 42, 206, 210
  and partial cell functions, 3
  and side chain theory, 9-10
Einstein, Albert, 146, 215
electrons, 39, 43, 47-48, 144, 148
electroporation, 229, 288
embryogenesis, 25, 30-31, 35, 108, 135, 167, 173
  and wound healing, 174-76, 182

embryonic stem cells (ESCs), 32–34, 174
Encyclopedia of DNA Elements (ENCODE), 92–94
energy
   biological, 52, 81, 141, 144–45
   chemical, 20, 141, 143–44
enhancers, 93, 123, 166
epidermal growth factor receptor (EGFR), 135, 191
epigenetic modifications, 69, 133, 135, 166, 182
epithelial-to-mesenchymal transition (EMT), 262
epitope, 18
Epstein-Barr Virus (EBV), 82, 107
Eshhar, Zelig, 232
evolution
   and genetic variation, 46, 51
   of life on earth, 100
   and selective pressure, 52, 87, 188
   of tumors, 130–31, 137
   of viruses, 106
exosomes, 177-78, 278
exotoxin, 254-55, 257
   exotoxin A, 254
extracellular matrix (ECM), 40, 174, 225, 262, 278
   cytoskeleton, 40

F
free radicals, *48*, 49
   See also reactive oxygen species
Folkman, Judah, 155–56

G
gatekeeper functions, 56, 59, 132, 187
   damage to, 117, 195
gene editing, 267–70
genetic code, 43, 45, 49, 71, 76–78, 267
genetic instability, 123, 129–30, 182, 188, 195, 229
germ cells, 110, 164, 267
germline mutation, 57
glucose fermentation, 147, 152
glutamine, 145, 266
glycolysis, 143, 152
   aerobic, 148
granzymes, 212

H
Hallmarks of Cancer, 180, *181*
Hanahan, Douglas, 179–81
hematopoiesis, 32, *33*
hematopoietic stem cells, *33*, 119, 263
hepatitis B virus, 61, 107, 250
Herceptin, 20, 191, 197, 261
histocompatibility locus antigen (HLA), 22, 218, 228–30
HLA restriction, 228, 232, 269

Index

Honjo, Tasuku, 243
Human Genome Project (HGP), 81-83, 91, 227
human papilloma virus (HPV), 61, 107, 250
Hungerford, David, 120–22
hybridoma, 17, 58, 216
hypoxia, 151–53, 183, 188
hypoxia-inducible factors (HIFs), 151–53

I
inflammation, 48, 160-163
  acute, 21
  and atherosclerosis, 160
  autoimmunity and, 107, 189
  chronic, 21, 61, 107-8, 112, 183, 188
  definition of, 20-21
  smoldering, 107, 149
immune checkpoint inhibitors, 243–44, 259, 265, 270, 274, 277
immunoediting, 247
immunoselection, 23-24
immunosurveillance, 23, 182, 230
immunotherapy. *See* cancer immunotherapy
immunotoxins, 254–57, 263
  innate immunity, 7-8, 22, 33, 224, 275
interferons, 97–98, 245

J
Jacob, Francois, 73, 94
Jenner, Edward, 2–3
Jerne, Niels, 10, 17
June, Carl H., 261, 264
  and CAR-T therapy, 235-39
  and engineered TCR., 231-33
  training of, 230-31
Junk Hypothesis, 80-81, 92-93

K
karyotype, 122
Knudson, Alfred, 56–58
Koch, Robert, 1-2, 6
Kohler, George, 16–18, 58, 216, 232
Krebs, Sir Hans, 144
  Krebs Cycle, 144, 146-47, 266
Kymriah, 237

L
leukemia, 20, 160, 197, 230, 236–37, 255
  acute lymphoblastic leukemia (ALL), 253
  chronic lymphocytic leukemia (CLL), 237
  chronic myelogenous leukemia (CML), 119, 260
  hairy cell, 255
leukocytes, 27, 160
liquid biopsy, 271
lymphoblast, *11*, 12

lymphocytes, 5, 8, 10, 17, 208, 217, 225
lymphoma, 20, 197, 211, 237, 256, 280
   Burkitt's lymphoma, 107
   non-Hodgkins, 20
lysosome, 22, 138, 149, 256

M

macrophages, 8, 23, 177, 224-26, 246, 251
   polarization of, 224-25
macropinocytosis, 149
Major Histocompatibility Complex (MHC), 22, 219, 230, 242
mass spectrometry, 220
matrix metalloproteinases (MMPs), 176, 195, 225, 262, 278
mechanism of action, 4
mechanistic pillars of cancer, 181-83, 265
Mechnikov, Ilya, 5–7
melanoma, 133, 155, 212, 220, 226, 243, 249, 273
Mello, Craig, 95–97
memory B cell, *11*, 12
Meselson, Matthew, 94
mesenchymal-to-epithelial transition (MET), 175
messenger RNA (mRNA), 73–78
   exosomes and, 177
   processing of, 100
   transcription of, 99
   vaccines containing, 250, 252
metabolism, 47–48, 94, 169, 261, 268
metastasis, 182, 186, 190, 194, 262
   emboli and, 172
   epigenetics and, 69
   genetic damage and, 130, 167, 182
   metastatic niches, 178, 188
   tumor cell migration, 176
   word origin of, 171
microbiome, 275–78
   dysbiosis of, 276
   fetal microbiota transplant (FMT), 277
microtubules, 256
mitosis
   description of, 52, *53*, 54
   and microtubules, 52
   and p53, 62
   signaling of, 133
   and spindle fibers, 256
molecular mass, 40
Monod, Jacques, 73
moxetumomab pasudotox, 255
Mukherjee, Siddhartha, xvi, 56–57
Mullis, Kary, 84, 88–89
mutator phenotype, 118, 129–

30
myeloid-derived stromal cells (MDSCs), 251

N
National Cancer Act, 60
National Cancer Institute (NCI), 117, 157, 205
  Rosenberg lab at, 210, 235
natural killer (NK) cells, 21, *33*, 177, 224
neoantigens, 219–22, 227–29, 246-47, 251
  mutation-derived, 220
  potential, 228
  targeted, 230
netosis, 161
neutrophils, 22, 161, 226, 246
  Neutrophil Extracellular Traps (NETs), 161
Nirenberg, Marshall, 76
Nixon, Richard, 60
non-coding RNAs, 93, 99
  See also RNA transcripts
nonsynonymous mutations, 221
Nowell, Peter, 122
nucleosomes, *66*, *67*
nucleotide bases, 43, 45, 51, 74, *142*
Nurse, Paul, 55

O
objective response rate (ORR), 226, 249
Ohno, Susumu, 79–81
Ohsumi, Yoshinori, 138
oncofetal antigens, 219
oncogenes, 107–8
  activation of, 110, 120–21, 129, 162, 187
  discovery of, 105
  oncogene addiction, 191
  protein products of, 107, 109, 132, 135
  signaling by, 149, 266
oncogenesis, 59, 224
oncolytic viruses, 264, 272–74
  activation of immune response by, 247
  definition of, 244
  development of, 249
  efficacy of, 259
organelles, 139, 193, 247, 275
  autophagy of, 138, 149
  definition of, 53-54
  mitochondria, 53, 139, 141, 143-48, 151-53
  mitochondrial DNA (mtDNA), 144-45
osteosarcoma, 60
overall survival (OS), 197
oxidation-reduction reactions, 144
oxidative phosphorylation, 141, 144, 147–48, 152

P

p53 tumor suppressor, 63, 110
   discovery of, 62
   functions of, 126, 136–37, 139
   inactivation of, 63
Paget, Stephen, 172
   and seed and soil hypothesis, 173
passenger mutations, 117–18
Pastan, Ira, 254
Pasteur, Louis, 1
PD-1, *234*, 243-44, 265, 270, 277
peptide bond, 41, *42*, 43, 152
perforin, 212
P-glycoprotein, 192-93
phagocytosis, 6, 149, 193
   definition of, 6
   phagocytes, 6, 22, 161
Philadelphia chromosome, 119-20, *121*, 129, 260
polymerase chain reaction (PCR), 88–89
precision medicine, 260
   precision oncology, 260
Prehn, Richmond, 213, 218
primers, 85-6
promoters, 93
   and gene activation, 107, 123
   and hypoxia-responsive elements, 151–52
proteasome, 152

protein-coding genes, 79–80, 91–92, 117
protein domains
   co-stimulatory, 236
   extracellular, 134
   intracellular, 134
   transmembrane, 134, 143, 236, 243
   See also antibodies, domains of
protein kinases, 119, 134
   cyclin-dependent kinases, 55
protein sequence, 72–73, 77-78, 213, 221
protein structure, 43, 69, 108
   prediction of, 44
   structure of amino acids, 43
   structure of antibodies, 13, *14*, 15
protein synthesis, 53, 73, 77, 94, 96, 99, 152
   exotoxin and, 254
   in herpes viruses, 245
proteome, 247, 260
proto-oncogenes
   activation of, 110, 120, 129, 132, 162
   definition of, 108
   functions of, 108–9
protozoans, 41, 275

Q

quality control systems, cellular, 47, 50, 110–11, 138

cell cycle controls, 55
degradation of, 20
DNA damage controls, 116
quiescence, 55, 62, 178

R
rabies virus, 206
radiation therapy, 186, 225
  damage from, 194
  targeted, 278
random mutations, 80, 115
reactive oxygen species (ROS), *48*, 110, 137, 143, 153, 161-62, 188
receptors, 41
  desensitization of, 224
  EGF receptor, 135, 191
  HER2 receptor, 191, 261
  ligand binding by, 134
  overstimulation of, 224
  and receptor-mediated endocytosis, 193
  structure of, 134
  targeting of, 178, 190
recombinant DNA, 12–13, 214, 268
Reinherz, Ellis L., 214-18
replicative exhaustion. *See* cellular senescence
retinoblastoma, 56–60, 111, 136
retroviruses, 97, 106
  human immunodeficiency virus (HIV), 88, 97, 106, 216, 231
ribosomal RNAs, 73
ribosome, 73, *74,* 75, 94
ribozymes, 100
Rituxan, 197
RNA Inducible Silencing Complex (RISC), 98
RNA interference
  discovery of, 99
  mechanism of, 96
  small interfering RNAs (siRNAs), 98–99
RNA polymerases, 85, 92, 101
RNA transcription, 72-73, 100
  microRNAs, 99, 169, 175, 177, 182, 267-68
  non-coding RNAs, 175
RNA World, 101
Rosenberg, Steven A., 209-10
  CAR-T research by, 228, 235–36
  tumor-infiltrating lymphocytes research by, 211-12, 223–27
Rous, Peyton, 103–5, 109

S
Sabin, Albert, 250
Sanger, Fred, 82
sarcomas, 175
  Coley's Toxin, treatment by, 200–204
  osteosarcoma, 60, 114

Rous sarcoma virus (RSV), 104–5
SARS-CoV-2, 88
Schleiden, Matthias, 3, 159
Schlossman, Stuart, 214, 216
Schwann, Theodor, 3, 159
selective pressure, 52, 87, 188, 195
single-chain variable fragment (scFv), *235*, 236
single nucleotide polymorphisms (SNPs), 84, 93
Sinsheimer, Robert, 84
sites, antigenic, 252
solid tumors, 132, 243, 257
  and delivery of therapeutics, 278
  and oncolytic viruses, 249
  treatment of, 249, 273
somatic cell nuclear transfer (SCNT), 164–65, 169
somatic cells, 163–64, 268
  differentiated, 163
  mutation of, 57
  normal human, 126
somatic mutation, 57, 116
somatic mutation theory of cancer, 116
spindle fibers, 54, 256
stem cells
  embryonic. *See* embryonic stem cells
  hematopoietic, *33*

multipotent, 34
normal, 169–70, 196
pluripotent, 33-34, 165
quiescent cancer in, 146
stem cell niches, 196
stem cell theory of cancer, 168
Stevens, Leroy, 25–31, 36
stimulation
  antigenic, 270
  biochemical, 224
  electrochemical, 34
  immune, 204
stress
  environmental, 46–47, 51, 84, 122, 168
  extreme, 136
  hypoxic, 154
  unrelenting, 131
stromal cells, 178, 197, 278
  bystander, 194
  myeloid-derived, 251
subcellular compartments, 72
sugar-phosphate backbone of DNA, *70*, 71, 79
Sulston, John E., 95
superoxide, *48*, 153
surface proteins, 11, 41, 173, 220, 232, 252, 270
  as receptors, 178, 190
  targeted, 256
syngeneic, 210, 212–13
synonymous mutation, 221

T

Talimogene laherparepvec (TVEC), 244–45, 247–49
Taq polymerase, 87
T cell receptor (TCR), 228, 231-32, 253, 269
T cells, 8, 22
  anti-tumor properties of, 208–10
  cytotoxic, 212, 216
  co-stimulation and, 232-33, *234*, 235-36, 241-42, 269-70
  maturation of, 217
  T cell exhaustion, 223, 270
  T regulatory cell, 224
telomeres, 125–26
  dysfunction of, 127
  lengthened, 127
  shortened, 126
Temin, Howard, 105
teratomas, 27, 36, 110
three-dimensional structure, 69, 86, 166
tissue remodeling, 190, 225
Tomasetti, Cristain, 114-15
totipotent cells, 32, 35
transcription factors, 166
  binding sites for, 79, 92–93
  cancer, role in, 182, 190
  definition of, 63
  EMT, role in, 174–75
  non-coding RNAs,
    interactions with, 99
  reprogramming of cells with, 165-167, 265
transcriptome, 260
transfer RNA (tRNA), *75*, 76, 78, 92, 100
translocation, 119-20, 122-23, 129
transplantation, 29
  allogeneic transplant, 270
  autologous transplant, 206, 223
  bone marrow transplant, 230, 247
transporters, 143, 152
  common ABC transporter, 192-93
tricarboxylic acid cycle (TCA cycle), 144
  See also Kreb's cycle
tumor antigens, 223, 228, 238, 246, 269
  patient's, 228
tumor cells
  proliferation of, 152, 156
  transplantation of, 29, 177
  treatment-resistant, 170
  unstable, 265
tumor development. *See* carcinogenesis
tumor dormancy, 178
tumor eradication, 265, 272
tumor escape, 269

tumor formation. *See* carcinogenesis
tumor growth. *See* tumor progression
tumorigenesis. *See* carcinogenesis
tumor-infiltrating lymphocytes (TILs), 211, 221, 227-28, 273
tumor initiation. *See* carcinogenesis
tumor lysis syndrome, 189
tumor microenvironment (TME), 168, 195-96
   characteristics of, 149, 168, 188, 194
   immunosuppression by, 221, 238, 250, 274
   microbiome in, 276
   pro-inflammatory properties of, 194, 278
   signaling in, 177, 195
   support of tumor cell resistance by, 190
tumor progression, 186, 262
   co-option of normal cells and, 132, 190
   differentiation and, 169
   DNA damage and, 116, 130
   epigenetics and, 69
   exosomes and, 177-78
   hypoxia and, 153
   immunity and, 208
   microbiome and, 276
   oncogenes and, 105, 107
   p53 protein and, 64
   viruses and, 107
tumor recurrence, 205
tumor regression, 200–201, 206–7
   complete, 211
   spontaneous, 199–200, 207
tumor suppressors, 63, 132, 136, 139, 149, 162, 187, 266
   function of, 59, 63, 123
   mutated, 266-67
two-mutation hypothesis, 58
tyrosine kinase, 119

U

ubiquitination, 151-52

V

vaccines, 250–51
   cancer vaccines, 251, 259, 273
   DNA vaccines, 251-52
   for measles, 207
   for polio, 250
   prophylactic, 250
variolation, 2–3
Varmus, Harold, 105
vascular leak syndrome, 257, 264
viral infections, 248
   cancer and, 61, 107
   inflammation from, 112, 162
   interferon discovery and, 96–97

oncogene discovery and, 106
p53 discovery and, 62
resistance to, 246
therapeutic use of, 206, 245
viral interference and, 97
viral interference, 97
viral proteins, 62, 246, 250
Virchow, Rudolf Carl, 160, 172
Vogelstein, Bert, 113
Vogt, Peter, 105
von Behring, Emil, 6-7, 9

W

Warburg, Otto H., 143, 146
Warburg Effect, 148, 153
Watson, James, 36, 47, 71, 81
Weinberg, Robert A, 179-81
Weismann Barrier, 163
Whitehead, Emily, 264
Wilmut, Sir Ian, 165
wound healing, 160–62, 174–76, 182
wound repair, 162, 167, 173, 176, 190, 195

X

x-ray crystallography, 76

Y

Yescarta, 237

Z

zygote, 25, 32, 44

About the Author

Drew Nathaniel Kelner, Ph.D., is a biochemist and immunologist with 35 years of experience in the biopharmaceutical industry, including twelve years at Amgen, the world's largest independent biotechnology company. During his career, he has been engaged in the development of innovative medicines for cancer, hemophilia, anemia, osteoporosis, and rheumatoid arthritis that have been used by millions of people around the world. A Long Island native who was educated at Haverford College (B.S., Chemistry) and Duke University (Ph.D., Biochemistry), he lives in Charlottesville, VA, where he is a loving husband and father and the president of Shenandoah Biotechnology Consulting, LLC. This book is the product of a long and exhilarating career in the scientific trenches of the biopharmaceutical industry and an homage to his lifelong preoccupation with experimental biology.

Printed in Great Britain
by Amazon